Classworks
Numeracy

Series editors

Len and Anne Frobisher

Len Frobisher,

John Taylor, John Spooner, Thelma Page

Ray Steele, Mike Spooner, Anitra Vickery

Contents

Introduction

How *Classworks* works

What this book contains

- Visual resources for structuring mental, written and problem-solving work.
- Examples of modelled mathematical methods and solutions.
- Lesson ideas including key questions and plenary support.
- Photocopiable pages to aid and structure pupil work.
- Blocked units to slot into medium-term planning.
- Oral/mental starter ideas to complement the daily teaching of mental facts and skills.
- Every idea is brief, to the point, and on one page.

How this book is organised

- There are blocked units of work from one week to several, depending on the strand.
- Each blocked unit is organised into a series of chunks of teaching content.
- Each 'chunk' has accompanying suggestions for visual modelling of teaching.
- For many teaching ideas we supply photocopiable resources.
- The objectives covered in the units are based on DfES sample medium-term planning.
- The units are organised in strand-based chunks, in a suggested order for teaching.

Planning a unit of work

How to incorporate *Classworks* material into your medium-term plan

- Pick the most relevant unit for what you want to teach – the units are organised in strands, sequentially according to the DfES sample medium-term plans.
- To find the content, look at the objectives on the first page of every unit.
- Or just browse through by topic, picking out the ideas you want to adapt.
- Every page has its content clearly signalled so you can pick and choose.
- Choose a generic starter from the bank at the back of the book, if required.

What each page does

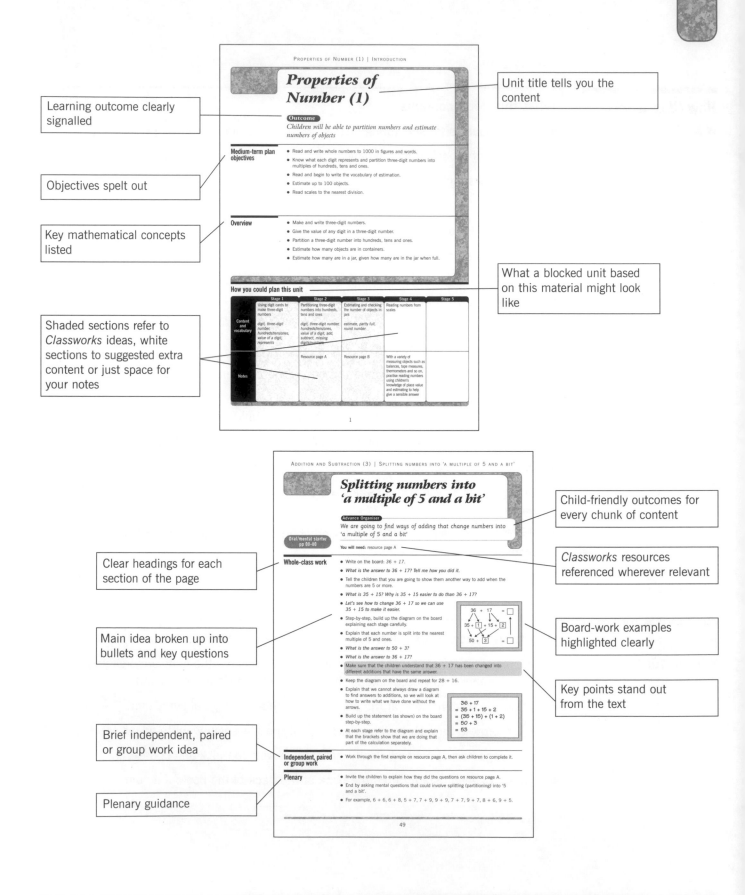

Learning outcome clearly signalled

Objectives spelt out

Key mathematical concepts listed

Shaded sections refer to *Classworks* ideas, white sections to suggested extra content or just space for your notes

Unit title tells you the content

What a blocked unit based on this material might look like

Clear headings for each section of the page

Main idea broken up into bullets and key questions

Brief independent, paired or group work idea

Plenary guidance

Child-friendly outcomes for every chunk of content

Classworks resources referenced wherever relevant

Board-work examples highlighted clearly

Key points stand out from the text

Properties of Number (1)

Outcome

Children will be able to partition numbers and estimate numbers of objects

Medium-term plan objectives

- Read and write whole numbers to 1000 in figures and words.
- Know what each digit represents and partition three-digit numbers into multiples of hundreds, tens and ones.
- Read and begin to write the vocabulary of estimation.
- Estimate up to 100 objects.
- Read scales to the nearest division.

Overview

- Make and write three-digit numbers.
- Give the value of any digit in a three-digit number.
- Partition a three-digit number into hundreds, tens and ones.
- Estimate how many objects are in containers.
- Estimate how many are in a jar, given how many are in the jar when full.

How you could plan this unit

	Stage 1	Stage 2	Stage 3	Stage 4	Stage 5
Content and vocabulary	Using digit cards to make three-digit numbers *digit, three-digit number, hundreds/tens/ones, value of a digit, represents*	Partitioning three-digit numbers into hundreds, tens and ones *digit, three-digit number, hundreds/tens/ones, value of a digit, add, subtract, missing digits/numbers*	Estimating and checking the number of objects in jars *estimate, partly full, round number*	Reading numbers from scales	
Notes		Resource page A	Resource page B	With a variety of measuring objects such as balances, tape measures, thermometers and so on, practise reading numbers using the children's knowledge of place value and estimating to help give a sensible answer	

Using digit cards to make three-digit numbers

Oral/mental starter p 183

Advance Organiser

We are going to make some three-digit numbers

You will need: 1 to 9 digit cards (for you), place-value arrow cards (per child)

Whole-class work

- From the digit cards 1 to 9 choose any three, say, 2, 4 and 7.

- Hold them up separately.

- *Tell me a three-digit number I can make with these digits.*

| 2 | 4 | 7 |

- Record the number in figures using boxes to represent the cards.

- Repeat until all six possible numbers have been made.

- You may wish to include 0 as one of the digits as this provokes discussion about 0 in the hundreds place of a three-digit number.

- In a random order, point to each number.

- *What is this number? How many hundreds has this number? How many tens has this number? How many ones has this number?*

- Give each child a set of hundreds, tens and ones place-value arrow cards.

- Ask the children to show one of the six numbers that has 7 tens.

- Repeat for other hundreds, tens and ones values.

Independent, paired or group work

- Ask the children to use the digits 1 to 9 once each to write three lots of three-digit numbers. They write each number in words, then write the value of each digit as it appears in their number.

Plenary

- Say *four hundred and thirty-six.*

- Ask for the value of the largest digit in the number.

- Repeat with other three-digit numbers, sometimes asking for the smallest digit's value.

- End by asking the children to use their number cards to make a number, say, that has 4 tens.

- Repeat many times for other digit values.

Partitioning three-digit numbers into hundreds, tens and ones

Advance Organiser

We are going to split three-digit numbers into hundreds, tens and ones

Oral/mental starter p 183

You will need: place-value arrow cards, resource page A (one per child)

Whole-class work

- Demonstrate using the 500, 40 and 7 cards to make the number 547.

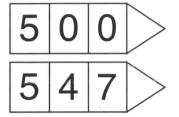

- *I am pulling out the 4 digit. What number will you see? What is the value of the 4 digit? How do you know?*

- Make 547 again. Repeat the questions for the 5 and the 7.

- Do this again for other three-digit numbers.

- Use the 200, 90 and 6 cards to make the number 296.

- *I am pulling out the 9 digit. What number will be left?*

- *What is 296 subtract 90? What is 90 add 206?*

- Make 296 again. Repeat for the 2 and the 6.

- Make 296 again and withdraw the digits in a different order.

- Do this again for other three-digit numbers.

- Use the 100, 80 and 3 cards to make the number 183.

- *When I pull the digits apart what three numbers will I see?*

- *What is 100 add 80 add 3?*

- Do this again for other three-digit numbers.

Independent, paired or group work

- Together, work through the first question on resource page A. Children can then complete the first three sections of resource page A.

- The last set of four questions are a challenge for the high achievers.

Plenary

- Together, go through resource page A.

- Ask the children to explain to the rest of the class how they worked them out.

Name: _____

Partitioning

Write in the missing digits.

364 = [ ] [0] [0] + [ ] [0] + [ ]

969 = [ ] [0] [0] + [ ] [0] + [ ]

305 = [ ] [0] [0] + [ ] [0] + [ ]

Write in the missing number.

571 = [ ] [ ] [ ] + 70 + 1

289 = [ ] + 200 + 80

733 = [ ] [ ] [ ] + 70 + 3

165 = [ ] [ ] [ ] + 60 + 5

Find the answer.

600 + 10 + 7 = [ ] [ ] [ ]

400 + 50 + 2 = [ ] [ ] [ ]

900 + 30 + 6 = [ ] [ ] [ ]

800 + 70 + 1 = [ ] [ ] [ ]

500 + 9 = [ ] [ ] [ ]   200 + 80 = [ ] [ ] [ ]

Estimating and checking the number of objects in jars

Advance Organiser

We are going to estimate the number of objects in jars

Oral/mental starter p 183

You will need: different transparent containers, a variety of small objects (marbles, beads, cubes), resource page B (one per child)

Whole-class work

- Have ready three different small transparent containers that contain objects up to 100 to the approximate levels of $\frac{1}{2}$, $\frac{1}{4}$ and $\frac{3}{4}$ of each container. Hold up the jar that is one-half full, say, of marbles.

- *How many marbles do you think there are in the jar? I will write some of your estimates on the board.*

- As you record estimates, discuss how sensible they are.

- Quickly count how many are in the jar. Compare the actual number with the estimates.

- Repeat the activity for the other two containers.

- Return to the container that is one-half full.

- Remind the children how many marbles were in the container, say 22.

- *What fraction of the container has marbles in it?*

- Record on the board: *The container is roughly half full. There are 22 marbles in about $\frac{1}{2}$ of the jar.*

- *Estimate the number of marbles that would be in the container if it were full.*

- Record some of the estimates, referring to doubling 22 to find how many. Stress that although you know exactly how many there are in the jar at the moment, you will still have to estimate how many in the full jar because the jar is roughly not exactly, half full.

- Explain that when estimating it is usual to round numbers, as it is impossible to estimate exactly.

- *Make another estimate, but this time make it a multiple of 5.*

- Record, and discuss, how sensible some of the estimates are in relation to the 22 marbles in half the container.

- Invite a child to put one marble at a time into the container, while everyone else counts.

- Compare the final number to the estimates and record it.

Independent, paired or group work

- Together, work the first question. Explain that by knowing how many are in the jar when it is full, helps to make a good estimate of how many are in the jar when it is partly full.

- Ask the children to complete resource page B.

Plenary

- Together, work through resource page B.

- Discuss the different methods that the children used to estimate the number in each jar, using the full amount.

Name: _____

Making estimates

When full the jar holds 30 marbles.

My estimate is _____ marbles.

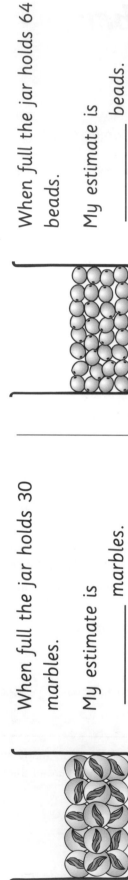

When full the jar holds 20 marbles.

My estimate is _____ marbles.

When full the jar holds 100 beads.

My estimate is _____ beads.

When full the jar holds 80 marbles.

My estimate is _____ marbles.

When full the jar holds 64 beads.

My estimate is _____ beads.

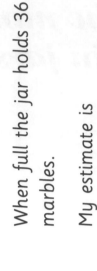

When full the jar holds 36 marbles.

My estimate is _____ marbles.

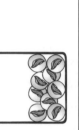

When full the jar holds 72 beads.

My estimate is _____ beads.

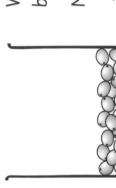

When full the jar holds 95 beads.

My estimate is _____ beads.

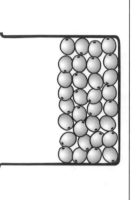

Classworks © Classworks Numeracy author team, Nelson Thornes Ltd, 2003

Properties of Number (2)

Outcome

Children will be able to use groupings to count objects, recognise odd and even two-digit numbers and use 'trial and improvement' to solve number puzzles

Medium-term plan objectives	• Count larger collections by grouping them in tens, then other numbers. • Count on or back in tens or hundreds, starting from any two- or three-digit number. • Count on or back in twos, starting from any two-digit number and recognise odd and even numbers to at least 100. • Solve number puzzles. • Explain methods and reasoning orally and in writing.
Overview	• Group objects in twos, threes, fives and tens. • Work out how many objects there are when grouped in twos, threes, fives and tens. • Continue even and odd sequences up to 100. • Recognise even and odd two-digit numbers. • Make general statements about the unit digits of even and odd numbers. • Use 'trial and improvement' to solve 'think of a number' puzzles. • Begin to explain the method and reasoning to solve number puzzles.

How you could plan this unit

	Stage 1	Stage 2	Stage 3	Stage 4	Stage 5
Content and vocabulary	Finding the total number in a set of objects by grouping *grouping, set*	Using number cards to make even and odd sequences *count on in twos, count back in twos, odd numbers, even numbers*	Solving 'think of a number' puzzles involving two numbers *puzzle, record in a table, trial and error, trial and improvement*		
Notes	Resource page A		Resource page B		

7

Finding the total number in a set of objects by grouping

Advance Organiser

We are going to find how many by grouping in different ways

Oral/mental starter p 183

You will need: small objects (buttons, counters, beads or small cubes), resource page A (one per child)

Whole-class work

- Give each pair of children between 30 and 40 buttons, counters, beads or small cubes.

- Ask them to split their objects into two sets with more in one than the other. One child has the larger set, the other the smaller one (tell them that will change next time).

- *Group your objects into sets of 5. Count how many you have in fives.*

- Ask them to record how many in each set as: ☐ groups of five and ☐ ones.

- Ask them to discuss with their partner how they could work out how many they have altogether, without moving the two sets, then to do it.

- They should record the total number in each set and how many altogether.

- Discuss the different strategies.

- Repeat, with the other child having the larger or smaller amount.

- The activity is then repeated, but with the objects grouped in twos, and finally in tens.

- Each time they should compare their answers with the answers they had using the different ways of grouping.

Independent, paired or group work

- Give each group of four children four copies of resource page A. Tell them to use one page to group the bees in twos to find how many. Then they should use the other pages, in turn, to group in threes, fives and finally tens. They must decide, in their group, who will do which grouping.

- At the end they have to write how many there are using the grouping and how many there are altogether.

Plenary

- Check how the children have written the totals using different groupings for resource page A.

- Ask the children to explain why the total is always the same whatever grouping is used.

PUPIL PAGE

Name: _____

Counting bees

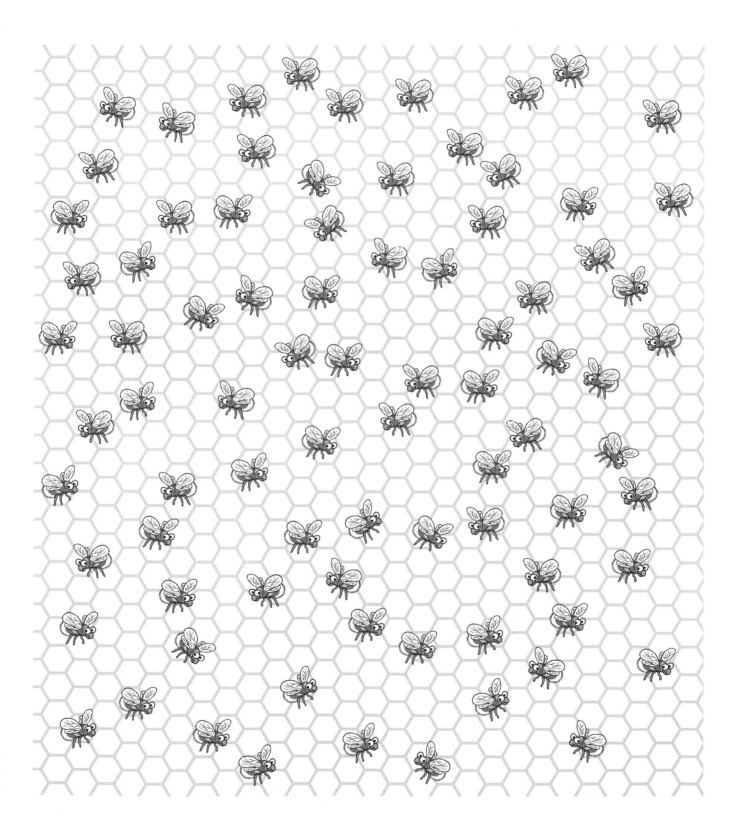

Using number cards to make even and odd sequences

Oral/mental starter
p 183

Advance Organiser

We are going to count in twos

You will need: number cards 50 to 100

Whole-class work

- Line up ten children in front of the class. Give the children the number cards 66 to 75 in a random order.

- *Stand in a line in order starting with 66.*

- Ask the rest of the class if there any numbers that are not in the correct place.

- Ask the children in the line to sort themselves into two lines: one line to be the odd numbers, the other to be the even numbers.

- Together, everyone says the even numbers in order from 66 to 74, then the odd numbers in order from 67 to 75.

- Ask the class to close and cover their eyes. Point to two children, one in each line, and ask them to turn their cards to show the blank side.

- Ask the rest of the class which even number is missing. Repeat for the missing odd number. Invite the children to continue the even and odd sequences.

- *Tell me what is special about the unit digits in each even number.*

- *Tell me what is special about the unit digits in each odd number.*

- Write on the board the two incomplete statements: *An even number ends in ....* *An odd number ends in ....*

- *Tell me how to complete each general statement.*

- Repeat the activity with a different set of ten children and a different set of ten consecutive numbers starting from any two-digit number between 50 and 90.

Independent, paired or group work

- Write some counting-on-in-twos and counting-back-in-twos sequences on the board. Ask the children to copy and complete each sequence. For example, 46, 48, 50..., 73, 75, 77..., 69, 67, 65..., 92, 90, 88....

Plenary

- Together, look at the sequences. Ask which numbers are even and which are odd.

Solving 'think of a number' puzzles involving two numbers

Advance Organiser

We are going to solve some number puzzles

Oral/mental starter p 183

You will need: resource page B (one per child)

Whole-class work

- Write on the board: *I am thinking of two numbers. The sum of the numbers is 15. The difference between the numbers is 3.*
- Draw a table on the board as shown below, without the numbers.
- Ask the children to make good guesses of what the two numbers could be.
- It is likely they will respond either with any two numbers, ignoring the two conditions, or by giving two numbers that have a sum of 15, ignoring the difference condition.
- Record each pair of numbers; for example, 2 and 7.
- Point to the first two numbers.
- *What is the sum of the two numbers?*
- Explain that the two numbers must have a sum of 15 and that this must be checked before working out the difference. Ask why this should be done.
- *Tell me two numbers that have a sum of 15.*
- Test each pair for the sum and the difference.
- Pairs that are very near, such as 10 and 5, can be used to develop the strategy of 'trial and improvement', whereas repeated wild guessing is 'trial and error'.

- The numbers 10 and 5 suggest changing the numbers slightly to retain the sum of 15, but change the difference from 5 to 3. Ask the children how this might be done.
- *We need to decrease the difference. We can do this by taking 1 off the 10 and adding 1 to the 5. This gives the two numbers, 9 and 6.*
- Record the 9 and 6, and show that they satisfy both conditions.

Two numbers			Sum of numbers	Difference of two numbers
2	and	7	9	5
10	and	5	15	5
9	and	6	15	3
	and			
	and			

Independent, paired or group work

- Ask the children to complete resource page B.

Plenary

- Invite the children to explain how they found the pair of numbers in resource page B.
- Look for explanations that show use of trial-and-improvement. Draw the children's attention to them, explaining why this is a good strategy.

(PUPIL PAGE)

Name: _____

Thinking of a number

I am thinking of two numbers.

The sum of the two numbers is 10.

The difference of the two numbers is 6.

What are my two numbers?

Two numbers	Sum of numbers	Difference of two numbers
☐ and ☐		
☐ and ☐		
☐ and ☐		
☐ and ☐		
☐ and ☐		
☐ and ☐		

Properties of Number (3)

Outcome

Children will be able to compare and order numbers, and round to the nearest ten

Medium-term plan objectives

- Read and write the vocabulary of comparing and ordering numbers, including ordinal numbers to 100.
- Compare two three-digit numbers and say which is more or less.
- Read and begin to write the vocabulary of approximation.
- Round any two-digit number to the nearest ten.
- Read scales and dials.

Overview

- Match ordinal numbers with ordinal names.
- Say and write the position of an object using ordinal numbers and words.
- Say which of two three-digit numbers is more and which is less.
- Say approximately the whole number an arrow is pointing to on an unmarked decade number line.

How you could plan this unit

	Stage 1	Stage 2	Stage 3	Stage 4	Stage 5
Content and vocabulary	Using ordinal number and word cards to position 'objects' *first, second (and so on), 1st, 2nd (and so on), position, ordinal, order*	Deciding which of two three-digit numbers is more/less *more/less, more/fewer, compare, is more/less than*	Positioning an arrow at a number on a decade number line *approximately, whole number, multiple of 10, arrow, position*	Reading scales and dials	
Notes	Resource page A		Resource page B	Ask the children to apply their knowledge of reading number lines to estimate unmarked 'between' numbers on measuring scales and dials	

13

Using ordinal number and word cards to position 'objects'

Oral/mental starter
p 183

Advance Organiser

We are going to use ordinal numbers and words

You will need: ordinal and ordinal name cards (1st to 20th and first to twentieth), resource page A (one per child)

Whole-class work

- Invite six children to stand in front of the class.
- Randomly give each child one of the ordinal cards, 1st to 6th.
- Ask the class: *Who is first?*
- Put a child in the 1st position in a line. Repeat for the 2nd to the 6th.
- In turn, point to any child in the line.
- *What position is Angela?*
- Repeat the activity with six other groups of children.
- Do the first activity again, but use the words *first* to *sixth*.
- Invite 12 children to the front of the class.
- Randomly give each of them either an ordinal number or word card from 1st to 6th.
- Ask them to find their partner.
- Tell them to exchange their card with someone else. Repeat the activity.
- Do this a number of times. Repeat the activity with numbers from first to twentieth.

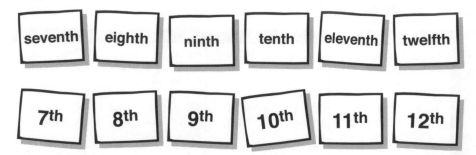

Independent, paired or group work

- Ask the children to complete resource page A.

Plenary

- Discuss resource page A with the children, asking them to explain their answers.
- Explain that an object can have a number that does not match its position.

(PUPIL PAGE)

Name: _____

Match partners

Classworks © Classworks Numeracy author team, Nelson Thornes Ltd, 2003

Deciding which of two three-digit numbers is more/less

Advance Organiser

We are going to find which of two numbers is more and which is less

Oral/mental starter p 183

Whole-class work

- Draw two boxes on the board. Write *347 packs* in one and *374 packs* in the other.

- *Which box has more/fewer packs?*

- On the board write:

  > 347 and 374
  > 347 is less than 374
  > 374 is more than 347

- Discuss the different ways children decided which was more/fewer.

- Explain that when using only numbers we use the word *less*. When referring to objects we use the word *fewer*.

- Tell the children the method for deciding which of two numbers is more.

- Write *374* and *347* in a place-value grid on the board.

H	T	U
3	4	7
3	7	4

- *Compare the hundreds digits. If they are equal, compare the tens digits. If they are the same, compare the units digits.*

- Discuss why the order is hundreds, tens and finally ones.

- Repeat with other pairs of three-digit numbers.

Independent, paired or group work

- Write pairs of three-digit numbers on the board.

- Ask the children to write which is more and which is less. They should write sentences such as *952 is more than 925; 925 is less than 952.*

Plenary

- Go through the children's sentences.

- *Which digits should we compare first? Which next? Who can explain why?*

16

Positioning an arrow at a number on a number line

Advance Organiser

We are going to find which number an arrow points to on a number line

Oral/mental starter
p 183

You will need: unmarked number line, 1 to 100 number cards per group, 1 to 9 cards per child, resource page B (one per child)

Whole-class work

- Show the children (or draw on the board) an unmarked number line with a box at each end.

- Write *0* and *10* in the boxes.

- Discuss the unmarked number line. Explain that the numbers 1 to 9 are missing, as are their positions. Tell them that the arrowheads at each end indicate that the line could be made longer and more numbers put on.

- *Draw an arrow where you think the number 5 should be. Tell me how you decided.*

- Repeat for other numbers from 1 to 9.

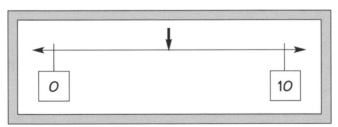

- Ask them to sort a set of 1 to 100 cards into 10s, 20s, and so on, and to put each decade in order with the smallest on top.

- Write *10* and *20* in the boxes on the number line. Draw the arrow at approximately 14.

- *The arrow is pointing approximately to a whole number. Hold up the number you think it could be. How did you decide?*

- Repeat for 20 to 30, 30 to 40, and so on.

- Repeat, writing *0* and *100* on the board, for multiples of 10.

Independent, paired or group work

- Ask the children to complete resource page B.

Plenary

- Together, look at the answers to questions on resource page B. Discuss the children's explanations about how they decided.

- Develop a strategy that decides on the numbers that are not shown, works out the middle or near middle position on the line, and matches to that the number that is half-way between the two known numbers. The position of other numbers can then be approximated from there.

Name: _____

Estimating numbers on a number line

Write in the whole number each arrow is pointing to.

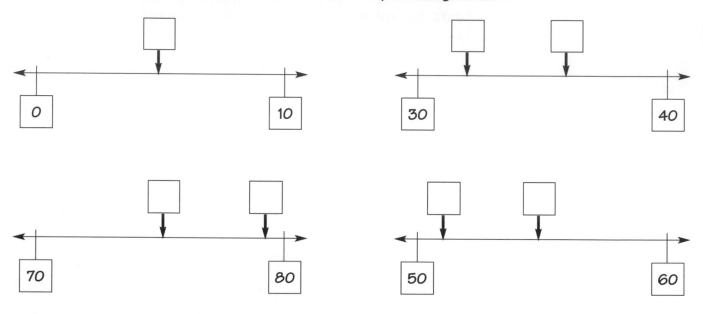

Write in the multiple of 10 that each arrow is pointing to.

The arrows are pointing to numbers that end in 5. Write what they could be.

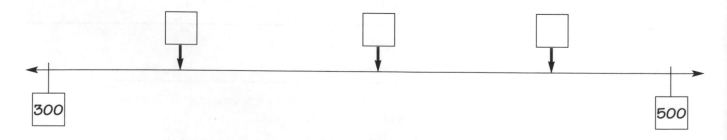

Properties of Number (4)

Outcome

Children will have investigated general statements and be able to count in sequences of 3, 4 and 5

Medium-term plan objectives

- Count on in steps of 3, 4 or 5 from any small number to at least 50 and back again.
- Create and describe simple number sequences.
- Investigate general statements about familiar numbers, and give examples that match them.
- Solve number puzzles.
- Explain methods and reasoning orally and in writing.

Overview

- Count on and back in threes, fours and fives.
- Write count-on and -back in threes, fours and fives sequences.
- State the rule for a sequence, given the sequence.
- Find examples to support a general statement.
- Find solutions to a number puzzle involving making totals the same.

How you could plan this unit

	Stage 1	Stage 2	Stage 3	Stage 4	Stage 5
Content and vocabulary	Practising counting in threes, fours and fives *start at, count on, count back, sequence, rule*	Finding examples that match and support a general statement *general statement, always true, proof, prove, examples to support*	Placing numbers 1 to 6 on diagrams to solve number puzzles *arrangement, totals, number puzzle*		
Notes	Resource page A		Resource page B		

Practising counting in threes, fours and fives

Advance Organiser

We are going to count in threes, fours and fives

Oral/mental starter p 183

You will need: resource page A (enlarged)

Whole-class work

- Use the enlarged 'Start at 0' to 'Start at 5' and 'Count on' cards from resource page A. Show one of each set to the children and put them into two piles.
- Split the class into groups of about five or six children. Make sure that the groups have a similar range of ability.
- Explain that you will turn over the top card of each pile and that will tell them a start number and a count-on number; for example, 'Start at 3' and 'Count on in 4s'.
- Each group then says the sequence together as quietly as possible until they reach a number greater than 50. They then stop.
- This is repeated two more times with other cards being turned over in pairs from the two piles.
- Organise a competition. Give each group a letter: A, B, C, and so on.
- Again, you have the two piles and turn over the top two cards.
- Group A says the sequence with each child saying a number in turn until a number greater than 50 is said.
- If a mistake is made, that child stands up. If on the next go the child says their number correctly, they sit down.
- Groups B, C and so on then do the same.
- When every group has said the sequence, the number of children standing is counted. Each group's score is recorded on the board.
- The game is played two more times. The final score is counted for each team. The winning team is the one with the least number.
- The competition is repeated using the 'Start at 46' to 'Start at 50' and the 'Count back in 3s, 4s, 5s' cards.

Independent, paired or group work

- Ask the children to complete some counting sequences, such as starting at 4 counting on in threes, starting at 51 counting back in fours, and so on.
- Write some sequences on the board and ask the children to continue them and write the rule.

Plenary

- Organise the order in which the class will say a sequence.
- Say: *Start at 3, Count on in 4s*. Point to a child to start. Each child says the next number according to the rule.
- Repeat for other rules using count on in twos, threes, fours, fives and tens.

| Start at 0 | Start at 1 | Start at 2 | Start at 3 | Start at 4 | Start at 5 |

| Start at 46 | Start at 47 | Start at 48 | Start at 49 | Start at 50 |

| Count on in 3s | Count on in 4s | Count on in 5s | Count back in 3s | Count back in 4s | Count back in 5s |

Finding examples that match and support a general statement

Advance Organiser

We are going to find examples that support a general statement

Oral/mental starter
p 183

Whole-class work

- Write on the board: *The sum of two multiples of 5 is a multiple of 5.*

- Remind the children that this is called a general statement. Discuss why it is called a general statement. (It applies to many, in fact an infinity, of numbers but does not state any particular number. It is general, not particular.)

- Draw a table on the board as shown. Discuss the table, questioning the children to make sure that they understand how the table helps to test the truth of the general statement.

  - *Tell me two multiples of 5.*

  - Say 20 and 35. Write the two numbers in the table.

  - *What is the sum of the two numbers? (55) Is 55 a multiple of 5? How can you tell?* Record 'Yes' in the table.

  - *The example supports the truth of the statement.*

The sum of two multiples of 5 is a multiple of 5		
Two multiples of 5	Sum	Is the sum a multiple of 5?
☐ and ☐	☐	
☐ and ☐	☐	
☐ and ☐	☐	

- Ask the children to work in pairs to find another two multiples of 5 and to test the statement. Collect evidence from the children to complete the table.

- Discuss that they have found examples, each of which *supports* the statement, but they have not proved the statement to be true.

- Tell the children that mathematicians have *proved* that the statement is always true. This means that there are no of multiples of 5 whose sum is not a multiple of 5.

Independent, paired or group work

- Ask the children to complete a similar table for the statement: *The difference between two numbers that end in 3 ends in 0.*

Plenary

- Collect the children's data about the general statement: *The difference between two numbers that end in 3 ends in 0* on the board. Stress that they have found examples that *support* the statement, but have not *proved* that the statement is always true.

- Ask the children to find an example that *proves* the general statement: 'The sum of two even numbers is an odd number' is not always true.

Placing numbers 1 to 6 on diagrams to solve number puzzles

Advance Organiser

We are going to solve a number puzzle

Oral/mental starter p 183

You will need: two sets of number cards 1 to 25, resource page B (one per child)

Whole-class work

- Draw an empty diagram on the board as shown below.

- Hold up the number cards 1 to 6. Explain that there are six boxes on the triangle and that one of the numbers 1 to 6 can be put in each box.

- Invite a child to tell you where to put each number.

- Point to the set of three numbers on each side of the triangle in turn.

- *What is the total of these three numbers?*

- Record each total using a number card in its shaded circle.

- *The three totals are different.*

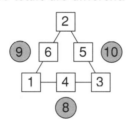

- Start again with a blank copy of resource page B.

- *The last arrangement of the numbers gave three totals that were all different.*

- Challenge the children to work in pairs to find an arrangement of the six numbers so that two of the totals are the same.

- Invite the children to explain their solution. One is shown here.

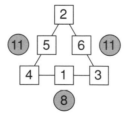

Independent, paired or group work

- Ask the children to complete resource page B. They try to find arrangements of the numbers 1 to 6 to satisfy the totals required. For each condition there are two different solutions to find.

Plenary

- Collect different solutions from the children for each possibility: all totals different, two totals the same and three totals the same. Discuss the children's strategies.

Name: _____

Solving a number problem

Make all three totals different.

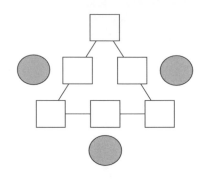   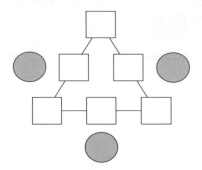

Make two totals the same and the third different.

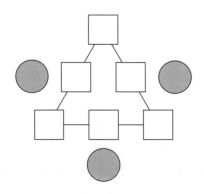   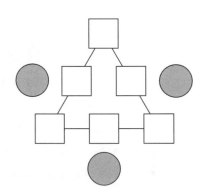

Make all three totals the same.

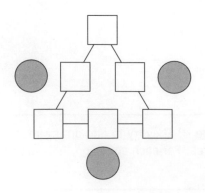   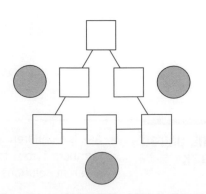

Classworks © Classworks Numeracy author team, Nelson Thornes Ltd, 2003

Properties of Number (5)

Outcome

Children will be able to use different strategies to order numbers

Medium-term plan objectives	<ul><li>Compare two three-digit numbers, say which is more or less and give a number that lies between them.</li><li>Round any three-digit number to the nearest 100.</li><li>Order a set of whole numbers to 1000 and position them on a number line.</li><li>Identify unlabelled divisions on a number line or measuring scale.</li></ul>
Overview	<ul><li>Find all numbers between two three-digit numbers.</li><li>Find the number that is halfway between two three-digit numbers.</li><li>Order a set of three-digit numbers.</li><li>Work out the value of a division on an unlabelled number line.</li><li>Find missing numbers on an unlabelled number line.</li></ul>

How you could plan this unit

	Stage 1	Stage 2	Stage 3	Stage 4	Stage 5
Content and vocabulary	Using number lines to find numbers between two numbers *between, halfway between*	Using a place-value grid to order three-digit numbers *order, smallest, largest, strategy*	Working out missing numbers on an unlabelled number line *divisions, sharing, value*		
Notes	Resource page A		Resource page B		

Using number lines to find numbers between two numbers

Advance Organiser

We are going to find numbers that come between two numbers

Oral/mental starter p 183

You will need: number cards 5, 6, 7, 50, 60, 70, 500, 600, 700, 2, 4, 72, 76, 432, 436, resource page A (one per child)

Whole-class work

- Draw a blank number line on the board as shown below. Remind the children that the arrows indicate that there are more numbers in both directions.

- Stick the number cards 5, 7, 50, 70, 500, 700 on the number lines as shown. Ask the children in what ways the number lines are alike and different.

- *Tell me a number that is between 5 and 7.* Ask where to record it. Put the number card 6 in its box.

- *Why is 6 halfway between 5 and 7?*

- Ask about more/less between the pairs 5 and 6, and 6 and 7. Repeat for 50 and 70, and 500 and 700. Point to the numbers 50 and 60. *Tell me a number that is between 50 and 60.*

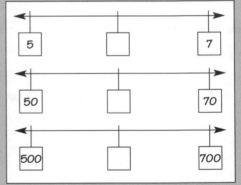

- *Show me approximately where the number will be on the number line. How did you decide?* Make a list of the numbers between 50 and 60. *Which number is halfway between 50 and 60?*

- Ring the number. Repeat for 60 and 70.

- Do the same for the multiples of 10 between 500 and 600, and 600 and 700.

- Point to the numbers 5 and 6. Ask the same questions as above. Repeat for numbers further apart.

- *Tell me the number that is halfway between 2 and 6.*

- *Why is 4 halfway between 2 and 6?*

- *Tell me the number that is halfway between 72 and 76.*

- *Tell me the number that is halfway between 432 and 436.*

- Record each number. Ask for numbers between any of the pairs of numbers.

Independent, paired or group work

- Ask the children to complete resource page A.

Plenary

- Together, work through the questions on resource page A.

- Collect different pairs of numbers for the solutions to the questions in the last section.

(PUPIL PAGE)

Name: _____

Halfway between

Write in the 'halfway between' numbers.

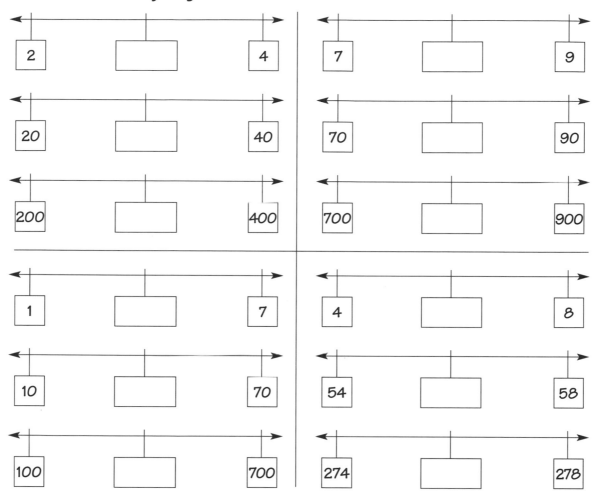

Write in the two numbers that the number shown is halfway between.

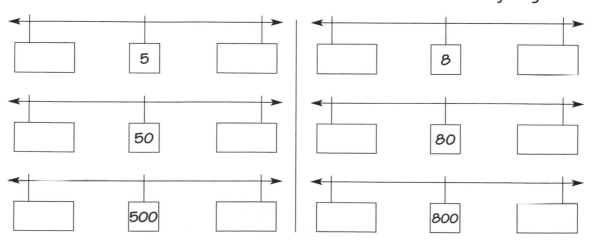

Classworks © Classworks Numeracy author team, Nelson Thornes Ltd, 2003

Using a place-value grid to order three-digit numbers

Oral/mental starter
p 183

Advance Organiser

We are going to put numbers in order of size

Whole-class work

- Write on the board: *518, 793, 276.*

- *Which is the largest number? Which is the smallest number?*

- Draw two sets of three boxes on the board with the words 'smallest' and 'largest' as shown.

- Ask the children to put the numbers in order, the smallest first.

- *How did you decide?*

- Repeat with largest first.

- Write on the board: *824, 831, 833, 822, 839.*

- Draw the grid below on the board. Ask the children to write each number in the grid.

- *Which is the largest number? Which is the smallest number? How did you decide?*

- Discuss the strategy of comparing hundreds, then tens, then ones.

- Draw on the board five boxes with the words 'smallest' and 'largest' at either end.

- *We are going to write these numbers in order, smallest first. How do we start?*

- *Which number has the smallest hundreds digit?*

- Discuss that because they all have the same hundreds digit, then we must look at the tens digits.

- Discuss that because more than one number has the same smallest tens digit, we must look next at their unit digits.

- Discuss why we have now found the smallest of the five numbers. Record the number in the first order box. Repeat, until the numbers are in order.

- Reverse the strategy to find the order with the largest first and compare the two orders.

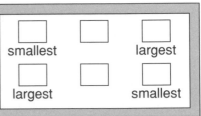

H	T	U
8	2	4
8	3	1
8	3	3
8	2	2
8	3	9

Independent, paired or group work

- Ask the children to order sets of five three-digit numbers, first transferring the numbers to a place-value grid and then ordering them, using the strategy.

Plenary

- Together, work through the examples. Repeatedly stress the strategy: 'compare hundreds, then tens, and finally ones'.

Working out missing numbers on an unlabelled number line

Advance Organiser

We are going to work out the numbers on a number line

Oral/mental starter p 183

You will need: resource page B

Whole-class work

- Draw a set of number lines on the board with two, four, five, eight and ten divisions as shown.

- Point to each number line in turn.

- *How many divisions has this number line?*

- You may need to explain what a division is. Many children think it is a mark, and they proceed to count the number of marks. Ask why it is called a division.

- On each number line, draw a jump line from the first mark to the next, and say: *This is a division on this line.*

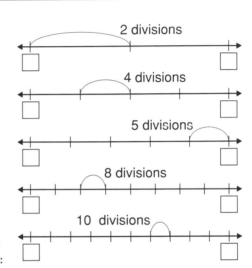

- On each number line in turn, point to each division and count how many altogether. Record above each line how many divisions the line has.

- Write in the numbers *0* and *40* in the boxes on the first number line. Make a jump with your finger from 0 to 40. *What is the value of this jump? How did you decide?*

- Make a jump with your finger from 0 of one division. *What is the value of this jump? How did you decide?*

- Point to the first mark after the 0. *What number is at this point?* Record the number.

- Repeat for the other number lines, recording the number at each division.

- Demonstrate with the children that to find the value of each division you need to find the difference between the numbers at each end of the line, then share that value between the number of divisions.

Independent, paired or group work

- Together, work out the missing number on the first number line on resource page B. Ask the children to complete the rest of the problems.

Plenary

- Establish the value of each division in each example in turn on resource page B.

- Check the answers by counting up in the value of a division starting at 0.

Name: _____

Number line divisions

Write in the missing numbers on each number line.

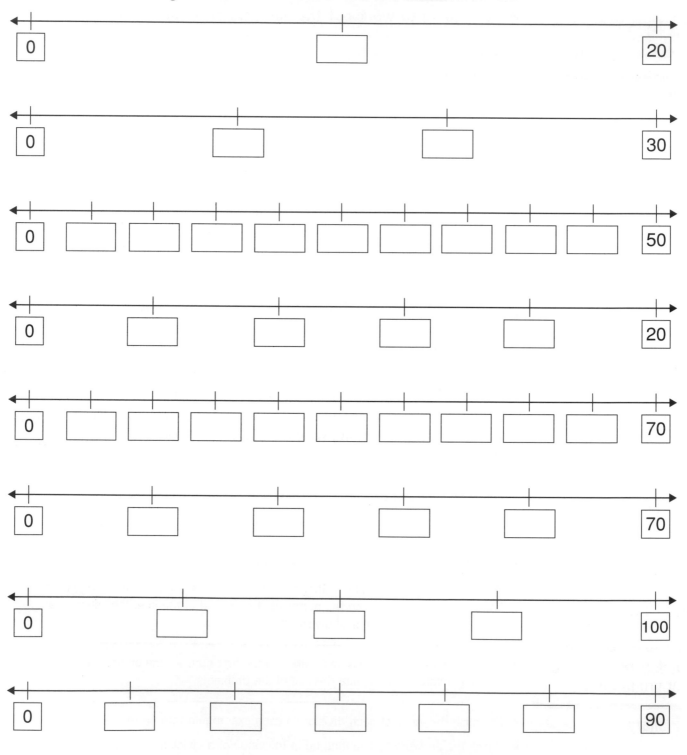

Properties of Number (6)

Outcome

Children will be able to recognise multiples of 2, 5, 10 and 50 and will have developed their number puzzle solving skills

Medium-term plan objectives

- Recognise two-digit and three-digit multiples of 2, 5 and 10, and three-digit multiples of 50.
- Solve number puzzles.
- Explain methods and reasoning orally and in writing.

Overview

- Say when a number is a multiple of 2, 5, 10 or 50.
- Say the general statement about end digits for the multiples of 2, 5, 10 and 50.
- Solve number puzzles involving making additions and subtractions.
- Solve a number puzzle involving making totals the same.

How you could plan this unit

	Stage 1	Stage 2	Stage 3	Stage 4	Stage 5
Content and vocabulary	Counting on to find multiples of 2, 5, 10 and 50 *count on in twos, fives, tens, fifties, multiple, pattern, end digit(s), general statement*	Investigating additions using the numbers 1 to 10 *addition, sets*	Placing numbers 1 to 6 on diagrams to solve number puzzles *arrangement, totals, number puzzle*	Investigating subtractions using numbers 1 to 10 *subtraction, sets*	
Notes	Resource page A	Resource page B	Resource page C	Stage 2 can be readily adapted to suit the investigation of subtractions	

Counting on to find multiples of 2, 5, 10 and 50

Advance Organiser

We are going to find multiples of numbers

Oral/mental starter
p 183

You will need: resource page A, 0 to 99 number grid

Whole-class work

- Together, count on in twos starting at 0. (Some children may find it helpful to have a 0 to 99 number square.)

- Draw some grids on the board as shown below.

- Point to the multiples-of-2 grid. Ask them to complete this, writing in order all the numbers that appear in a count-on-in-twos sequence that starts at 0.

- Remind them that these numbers are called multiples of 2.

- *Why are they called multiples of 2?*

- Let every child complete the grid.

- *What is special about the unit digit of a multiple of 2? How does the grid help you see the pattern?*

- Write on the board the general statement for multiples of 2: *A multiple of 2 ends in 0, 2, 4, 6 or 8.*

- Repeat the activity for multiples of 5 and 10.

- Together, count up in 50s from 0 to 1000: 0, 50, 100, 150, 200, 250 and so on to 950, 1000.

- Record the numbers in rows to establish the pattern in the last two digits:

0	*50*
100	*150*
200	*250*
300	*350 and so on.*

- *What is special about the last two digits of each multiple of 50?*

- Write on the board the general statement for multiples of 50: *A multiple of 50 ends in 00 or 50.*

Multiples of 2				Multiples of 5		Multiples of 10
0	2	4		0	5	0

Independent, paired or group work

- Ask the children to complete resource page A.

Plenary

- Together, check the numbers that should be ringed in each set. Remind the children about the general statements on multiples of 2, 5, 10 and 50.

- End by asking the children to tell you a number that is NOT a multiple of 2, 5, 10 or 50.

Name: _____

Finding multiples

Ring each multiple of 2.

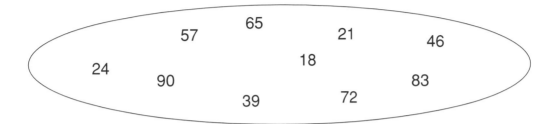

57   65   21   46
24   18
90   83
39   72

Ring each multiple of 5.

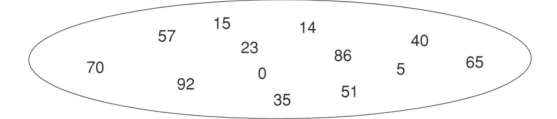

57   15   14   40
23   86   65
70   0   5
92   51
35

Ring each multiple of 10.

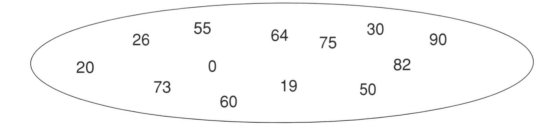

55   64   30   90
26   75
20   0   82
73   19   50
60

Ring each multiple of 50.

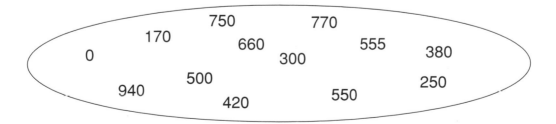

750   770
170   660   555   380
0   300
500   250
940   550
420

Classworks © Classworks Numeracy author team, Nelson Thornes Ltd, 2003

Investigating additions using the numbers 1 to 10

Advance Organiser

We are going to investigate making additions

**Oral/mental starter
p 183**

You will need: number cards 1 to 10 (one per child and one for you), resource page B (per pair)

Whole-class work

- Draw a blank addition on the board and show the children the number cards 1 to 10.
- Point out that there are three positions on the blank addition.
- *What should go in each box?*
- *Choose three cards that will complete the addition.* (Say, 2 + 3 = 5)
- Record the addition on the board. Put the three cards on one side.
- *There are seven numbers left. Choose three cards that will complete the addition.* (Say, 9 + 1 = 10.)
- Record as before.

$$\square \; + \; \square \; = \; \square$$

2 + 3 = 5
9 + 1 = 10

- *There are four numbers left. Can you choose three cards that will complete the addition correctly?*

Independent, paired or group work

- Children work in pairs to complete resource page B. Suggest that they use the numbers cards 1 to 10 and make jottings before recording their findings on the sheet.
- Attempting to make sets that work is not easy - there are a limited number of solutions. Encourage the children to be systematic; for example, starting with 1 + 9, 2 + $\square$, 3 + $\square$, or writing down number parts for numbers 1 to 10 first (10 = 1 + 9, 2 + 8 ...; 9 = 1 + 8, 2 + 7 ...).

Plenary

- Collect from the children the different possible solutions.
- Ask them how they found the different sets. Look out for interesting strategies to discuss.

Name: _____

Making additions

Here are ten numbers.

| 1 | 2 | 3 | 4 | 5 | 6 | 7 | 8 | 9 | 10 |

Use any nine of the ten numbers to make additions.
You cannot use a number more than once.

Investigate making as many different sets as you can

Set 1

☐ + ☐ = ☐

☐ + ☐ = ☐

☐ + ☐ = ☐

Set 2

☐ + ☐ = ☐

☐ + ☐ = ☐

☐ + ☐ = ☐

Set 3

☐ + ☐ = ☐

☐ + ☐ = ☐

☐ + ☐ = ☐

Set 4

☐ + ☐ = ☐

☐ + ☐ = ☐

☐ + ☐ = ☐

Classworks © Classworks Numeracy author team, Nelson Thornes Ltd, 2003

Placing numbers 1 to 6 on diagrams to solve number puzzles

Advance Organiser

We are going to solve a number puzzle

Oral/mental starter p 183

You will need: two sets of number cards 1 to 40, resource page C (one per child)

Whole-class work

- Draw a blank diagram on the board as shown here.

- Hold up the number cards 1 to 8.

- Explain that there are eight boxes on the square, and that one of the numbers 1 to 8 can be put in each box.

- Arrange the numbers 1 to 8 as shown.

- Point to the set of three numbers on each side of the square in turn.

- *What is the total of these three numbers?*

- Record each total using a number card in its shaded circle.

- *The four totals are different.*

- Start again with a blank diagram. Arrange the eight cards as shown.

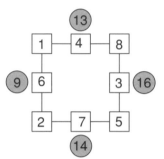

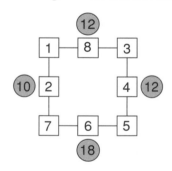

- *The last arrangement of the numbers gave three totals that were all different.*

- Challenge the children to work in pairs to find another arrangement of the eight numbers so that two of the totals are the same and two different.

- Invite the children to explain their solution.

Independent, paired or group work

- Ask the children to complete resource page C. They try to find arrangements of the numbers 1 to 8 to satisfy the totals required. For each condition there are two different solutions to find. They may find the last question more difficult than the others.

Plenary

- Collect different solutions from the children for each possibility: two totals the same, three totals the same and all four totals the same.

Name: _____

Solving number puzzles

Make two totals the same and two different.

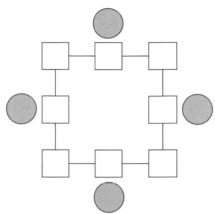   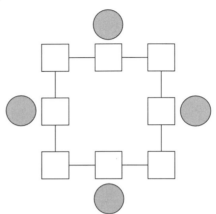

Make three totals the same and the fourth different.

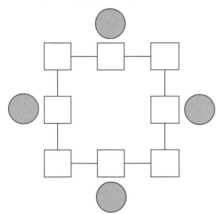   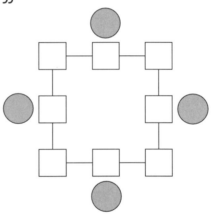

Make all four totals the same.

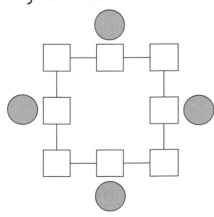   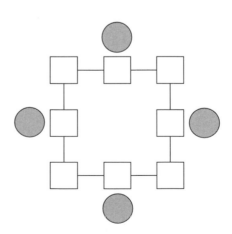

Classworks © Classworks Numeracy author team, Nelson Thornes Ltd, 2003

Addition and Subtraction (1)

Outcome

Children will be able to use near-doubles and partitioning

Medium-term plan objectives

- Extend understanding of the operations of addition and subtraction.
- Read and begin to write mathematical vocabulary.
- Use +, – and = signs.
- Recognise that addition can be done in any order.
- Put the larger number first in order to count on.
- Identify near-doubles.
- Bridge through a multiple of 10, and adjust.

Overview

- Relate near-doubles to an addition double.
- Use a double to find the answer to a near-double.
- Partition a single-digit number to subtract it from a two-digit number.

How you could plan this unit

	Stage 1	Stage 2	Stage 3	Stage 4	Stage 5
Content and vocabulary	Making families of near addition doubles *family, addition double, near-double*	Using diagrams to bridge 10 when subtracting numbers *split, partition, subtract*			
Notes	Resource page A	Resource page B			

Making families of near-addition-doubles

Oral/mental starter
pp 184–185

Advance Organiser

We are going to look for families of near-addition-doubles

You will need: resource page A (one per child and one enlarged)

Whole-class work

- Write some addition-doubles on the board, such as: *10 + 10, 25 + 25, 37 + 37.*

- *What is special about each of these additions?*

- Remind the children that they are called addition-doubles.

- *Why are they called addition doubles?*

- Ask for the answer to each, and how the children worked it out.

- Draw the following diagram on the board and write *20 + 20* in the central box.

- Add the following additions to the diagram: *21 + 20, 21 + 19, 20 + 19, 18 + 22.*

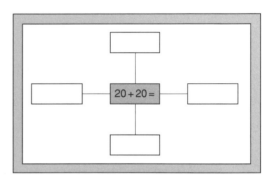

- *I am making a family of additions. Tell me why they make a family.*

- In turn, ask what the similarities and differences are between 20 + 20 and each of the other additions.

- Tell them that the family is called the 'near addition double family of 20 + 20'.

- *Why do you think it is called a 'near addition double family'?*

- Explain that we can find the answer to each addition in the family by using the answer to the addition-double.

- *What is the answer to 20 + 20?* Record on the board: *20 + 20 = 40.* Double 20 = 40.

- Explain that 21 + 20 is 1 more than 20 + 20.

- Record: *21 + 20 = double 20 plus 1 = 40 + 1 = 41.*

- Repeat for the other near-addition-doubles on the board.

Independent, paired or group work

- With an enlarged copy of resource page A, explain what the children have to do for each question.

- Ask them to complete the rest of the questions on the page.

Plenary

- Invite the children to describe and explain what they did for the questions on resource page A.

- Draw their attention to some near-addition-doubles being in more than one family.

- Show them that 28 + 29 = double 28 plus 1 = 56 + 1 = 57 and 28 + 29 = double 29 minus 1 = 58 – 1 = 57.

Name: _____

Near-addition-doubles

Write in the answer to 30 + 30 in the centre box.

Use this answer to work out the other near-doubles.

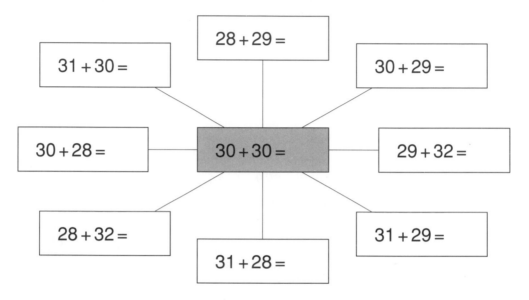

28 + 29 =

31 + 30 =

30 + 29 =

30 + 28 =

30 + 30 =

29 + 32 =

28 + 32 =

31 + 29 =

31 + 28 =

Write in the answer to 70 + 70 in the centre box.

Use this answer to work out the other near-doubles.

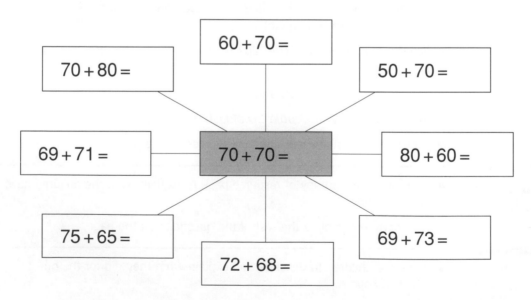

60 + 70 =

70 + 80 =

50 + 70 =

69 + 71 =

70 + 70 =

80 + 60 =

75 + 65 =

69 + 73 =

72 + 68 =

Classworks © Classworks Numeracy author team, Nelson Thornes Ltd, 2003

Using diagrams to bridge 10 when subtracting numbers

Advance Organiser

We are going to find a way of subtracting single-digit numbers

Oral/mental starter pp 184–185

You will need: resource page B (one per child)

Whole-class work

● Write on the board: *74 – 8*.

● *What is the answer to 74 – 8? Tell me how you did it.*

● Explain that you are going to show them a way to do the subtraction.

● Step-by-step, using questioning, build up the diagram on the board.

$$74 \quad - \quad 8 \quad = \quad \square$$
$$74 - 4 - 4$$
$$70 \quad - \quad 4 \quad = \quad \square$$

● *What is the answer to 70 – 4?*

● *What is the answer to 74 – 8?*

● Make sure the children understand that 74 – 8 has been changed into 70 – 4, but the answer is the same for both subtractions.

● Repeat for 56 – 9.

$$56 - 9 \quad = \quad \square$$
$$56 - 6 - 3$$
$$50 - 3 \quad = \quad \square$$

Independent, paired or group work

● Work through the first example on resource page B.

● Ask the children to complete resource page B.

Plenary

● Ask questions involving mental subtraction of a single-digit number from a two-digit number bridging 10, such as: 82 – 8, 63 – 9, 46 – 8, 94 – 6, 77 – 9, 55 – 7.

● Ask the children to explain to the class how they did them.

(PUPIL PAGE)

Name: _____

Bridging 10

Write in the missing numbers. Find the answers to each pair of subtractions.

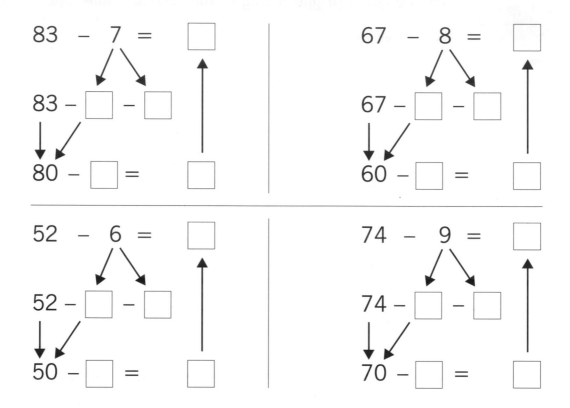

83 – 7 = ☐

83 – ☐ – ☐

80 – ☐ = ☐

67 – 8 = ☐

67 – ☐ – ☐

60 – ☐ = ☐

52 – 6 = ☐

52 – ☐ – ☐

50 – ☐ = ☐

74 – 9 = ☐

74 – ☐ – ☐

70 – ☐ = ☐

Work out the answers.

45 – 8 = ☐   81 – 3 = ☐

76 – 7 = ☐   74 – 5 = ☐

92 – 9 = ☐   57 – 9 = ☐

53 – 8 = ☐   61 – 9 = ☐

Classworks © Classworks Numeracy author team, Nelson Thornes Ltd, 2003

Addition and Subtraction (2)

Outcome

Children will understand subtraction as the inverse of addition and use open number lines to find small differences

Medium-term plan objectives

- Understand that subtraction is the inverse of addition.
- Say a subtraction statement equivalent to an addition statement and vice versa.
- Find a small difference by counting up from the smaller number.

Overview

- Explain why subtraction undoes addition.
- Find missing numbers on an inverse diagram.
- Use an open number line to find the difference between two close numbers.
- Find a small difference by counting on.

How you could plan this unit

	Stage 1	Stage 2	Stage 3	Stage 4	Stage 5
Content and vocabulary	Using diagrams to show how subtraction undoes addition *addition, subtraction, undoes, inverse*	Using open number lines to find small differences			
Notes	Resource page A				

43

Using diagrams to show how subtraction undoes addition

Advance Organiser

We are going to see how subtraction can undo addition

Oral/mental starter pp 184–185

You will need: linking cubes, resource page A (one per child)

Whole-class work

- Use linking cubes to make a 7-stick in red and a 5-stick in blue.

- Together, count how many cubes are in each stick. Join the two sticks.

- *How many cubes are there altogether? What is 7 + 5?*

- Record on the board: *7 + 5 = 12*. Show the combined stick. *How many cubes in this stick?*

- Take the 5-stick away. *How many cubes have I taken away? How many cubes are left?*

- Record *12 − 5 = 7* on the same line as *7 + 5 = 12*.

- *Tell me how the addition and the subtraction are the same and how they are different.*

- Put a double-headed arrow between them. *7 + 5 = 12 ↔ 12 − 5 = 7*

- Explain that subtraction undoes addition and we say that subtraction is the inverse of addition.

- Repeat with two other sticks.

- Draw on the board the following addition and subtraction diagram.

- Invite a child to tell you a number, say 9, and write it in the first box.

- Ask another child to tell you another number, say 8. Write this number in both ovals so it reads '+ 8' and '− 8'.

- *Work with a partner to find what the two missing numbers will be.*

- Ask what they notice about the numbers in the first and last boxes.

- Ask them why this is.

- Repeat the activity with two-digit numbers.

Independent, paired or group work

- Ask the children to complete resource page A.

Plenary

- Together, work through some of the questions on resource page A.

- Stress that subtraction undoes addition and that subtraction is the inverse of addition.

Name: _____

Find the missing numbers

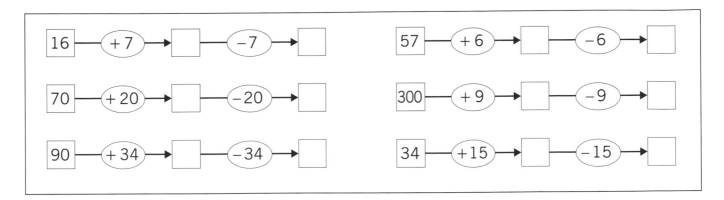

16 → +7 → ☐ → −7 → ☐

57 → +6 → ☐ → −6 → ☐

70 → +20 → ☐ → −20 → ☐

300 → +9 → ☐ → −9 → ☐

90 → +34 → ☐ → −34 → ☐

34 → +15 → ☐ → −15 → ☐

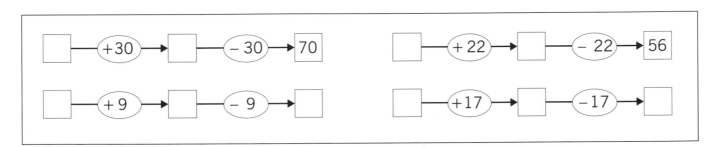

☐ → +30 → ☐ → −30 → 70

☐ → +22 → ☐ → −22 → 56

☐ → +9 → ☐ → −9 → ☐

☐ → +17 → ☐ → −17 → ☐

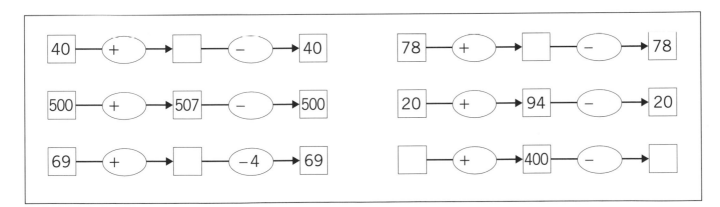

40 → + → ☐ → − → 40

78 → + → ☐ → − → 78

500 → + → 507 → − → 500

20 → + → 94 → − → 20

69 → + → ☐ → −4 → 69

☐ → + → 400 → − → ☐

Classworks © Classworks Numeracy author team, Nelson Thornes Ltd, 2003

Using open number lines to find small differences

Oral/mental starter
pp 184–185

Advance Organiser

We are going to find the difference between numbers that are very close to each other

You will need: open number lines

Whole-class work

- Write on the board: *What is the difference between 497 and 502?*

- Show the children an open number line. *This number line has no numbers on it.*

  ◄─────────────────────────►

- *How can we use it to find the difference between 497 and 502?*

- Explain that we can write 497 and 502 at either end of the open number line.

- *What is the multiple of 100 that is between 497 and 502?*

- Ask the children where, approximately, 500 is on the number line.

- Position 500 on the number line.

- *How many is it from 497 to 500?* Record the answer on the number line.

- *How many is it from 500 to 502?* Record the answer on the number line.

- *How many is it from 497 to 502?* Record the answer on the number line.

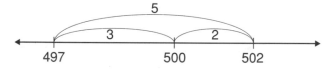

- Repeat for these pairs of numbers: 204 and 197, 663 and 595, 901 and 886.

- *What is the difference between 497 and 502?*

- Ask the children to close their eyes.

- *Make a picture in your minds of a number line. Put 497 and 502 where they should be. Where is 500?*

- *Count on from 497 to 500. Count on from 500 to 502. Altogether how many did you count on?*

- Repeat the activity with these number pairs: 303 and 298, 198 and 206, 395 and 402.

Independent, paired or group work

- Give the children similar pairs of numbers to find the difference between; for example, 765 and 758, 173 and 168, 496 and 504, 893 and 902 and so on. Some children will need the help of open number lines.

Plenary

- Discuss the children's answers.

- Stress that it is usually easier to position the smaller number first on the number line.

Addition and Subtraction (3)

Outcome

Children will be able to solve additions by reordering or partitioning mentally

Medium-term plan objectives

- Add three then four single-digit numbers mentally.

- Adding three or four small numbers by putting the largest first and/or finding pairs that total 10.

- Partition into '5 and a bit' to add 6, 7 or 8.

Overview

- Rearrange an addition of more than two numbers so that the largest number is first, and find the answer.

- Rearrange an addition of more than two numbers so that a pair that makes 10 is together and comes first, and find the answer.

- Add two two-digit numbers by splitting each number into 'a multiple of 5 and a bit'.

How you could plan this unit

	Stage 1	Stage 2	Stage 3	Stage 4	Stage 5
Content and vocabulary	Reordering additions with largest numbers first *larger/largest, pairs that make 10, rewrite, rearrange the order*	Splitting numbers into a multiple of '5 and a bit' *multiple of '5 and a bit', split numbers, partition*			
Notes		Resource page A			

Reordering additions with largest numbers first

Oral/mental starter pp 184–185

Advance Organiser

We are going to find easier ways to add more than two numbers

Whole-class work

- Write on the board: *2 + 7, 7 + 2.*
- *What is the answer to each addition? Why are the answers the same?*
- *Which is the easier addition to do? Why?*
- Explain that it is usually easier to count on from the larger number.
- Discuss why this is so.
- Write these additions on the board and ask for the answers: *1 + 6, 3 + 8, 2 + 9, 4 + 17.* Each time ask how the children did it.
- Stress that it is easier to count on from the larger number.
- Ask why this is.
- Write on the board: *8 + 6 + 2, 8 + 2 + 6.*
- *What is the answer to each addition? Why is the answer the same?*
- *Which is the easier addition to do? Why?*
- Explain that rearranging the numbers so that a pair that make 10 are next to each other makes the addition easier to do.
- Write these additions on the board and ask for the answers: *3 + 9 + 7, 4 + 6 + 12, 1 + 16 + 9.*
- Each time ask how they did it.
- Stress that making pairs that add to 10 makes the addition easier to do.
- Ask why this is.

Independent, paired or group work

- Ask the children to complete similar additions, repositioning the numbers as necessary.
- Early finishers can try harder examples such as 6 + 28 + 3, and sets of four numbers.

Plenary

- In turn, discuss each solution to the questions the children answered.
- Each time, point out that writing the largest number first and/or making pairs which add to 10 can make it easier to add.

Splitting numbers into a multiple of '5 and a bit'

Advance Organiser

We are going to find ways of adding that change numbers into a multiple of '5 and a bit'

Oral/mental starter pp 184–185

You will need: resource page A

Whole-class work

● Write on the board: *36 + 17*.

● *What is the answer to 36 + 17? Tell me how you did it.*

● Tell the children that you are going to show them another way to add when the numbers are 5 or more.

● *What is 35 + 15? Why is 35 + 15 easier to do than 36 + 17?*

● *Let's see how to change 36 + 17 so we can use 35 + 15 to make it easier.*

● Step-by-step, build up the diagram on the board explaining each stage carefully.

● Explain that each number is split into the nearest multiple of '5 and ones'.

● *What is the answer to 50 + 3?*

● *What is the answer to 36 + 17?*

● Make sure that the children understand that 36 + 17 has been changed into different additions that have the same answer.

● Keep the diagram on the board and repeat for 28 + 16.

● Explain that we cannot always draw a diagram to find answers to additions, so we will look at how to write what we have done, without the arrows.

● Build up the statement (as shown) on the board step-by-step.

● At each stage refer to the diagram and explain that the brackets show that we are doing that part of the calculation separately.

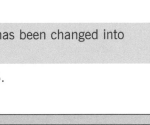

```
36 + 17
= 35 + 1 + 15 + 2
= (35 + 15) + (1 + 2)
= 50 + 3
= 53
```

Independent, paired or group work

● Work through the first example on resource page A, then ask the children to complete it.

Plenary

● Invite the children to explain how they did the questions on resource page A.

● End by asking mental questions that could involve splitting (partitioning) into '5 and a bit'.

● For example, 6 + 6, 6 + 8, 5 + 7, 7 + 9, 9 + 9, 7 + 7, 9 + 7, 8 + 6, 9 + 5.

Name: _____

'5 and a bit'

Write in the missing numbers.

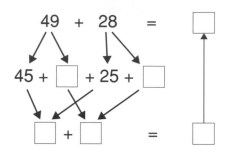

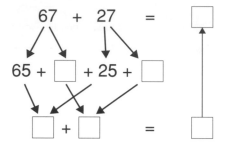

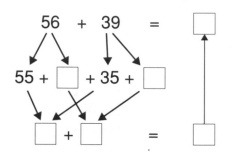

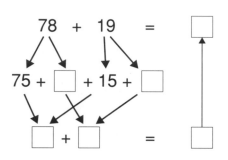

57 + 36 ☐

= 55 + ☐ + 35 + ☐

= (55 + 35) + (☐ + ☐)

= ☐ + ☐

= ☐

48 + 27 ☐

= 45 + ☐ + 25 + ☐

= (45 + 25) + (☐ + ☐)

= ☐ + ☐

= ☐

39 + 46 ☐

= 35 + ☐ + 45 + ☐

= (35 + 45) + (☐ + ☐)

= ☐ + ☐

= ☐

27 + 68 ☐

= ☐ + 2 + ☐ + 3

= (☐ + ☐) + (☐ + ☐)

= ☐ + ☐

= ☐

Addition and Subtraction (4)

Outcome

Children will be able to use base 10 apparatus to carry out additions of three-digit numbers and record this operation

Medium-term plan objectives

● Add three two-digit numbers using apparatus or informal methods.

● Partition into tens and units and recombine.

Overview

● Use base 10 apparatus to add up to three three-digit numbers.

● Partition two-digit numbers into tens and ones.

● Find the sum of two two-digit numbers without carrying, by partitioning into tens and ones.

How you could plan this unit

	Stage 1	Stage 2	Stage 3	Stage 4	Stage 5
Content and vocabulary	Using base 10 apparatus to add three-digit numbers *hundreds, tens and ones, combine*	Adding two-digit numbers by partitioning into tens and ones *partition, recombine*			
Notes	Resource page A				

51

Using base 10 apparatus to add three-digit numbers

Oral/mental starter
pp 184–185

Advance Organiser

We are going to use apparatus to find the sum of three-digit numbers

You will need: base 10 apparatus, resource page A (one per child)

Whole-class work

- Discuss the different shapes and their values in the base 10 apparatus.

- Record a picture of each piece and its value on the board.

- Invite three children to stand at the front of the class. One holds up one hundred, another four tens and the third, three ones.

- *Which number does the apparatus they are holding up represent?*

- Record the number on the board.

- Keep the children at the front. Repeat with the same material for 212.

- Record the 212 to make an addition with 143. Read out the addition.

- *How can we use the apparatus to find the sum of 143 and 212?*

- Discuss the different methods. Explain that we can combine the ones, then the tens and finally the hundreds. Do this together.

- *How many ones are there altogether? How many tens are there altogether?*

- *How many hundreds are there altogether? What is the answer to 143 + 212?*

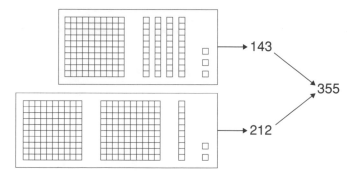

- Repeat the activity for the three amounts 221, 134 and 413.

Independent, paired or group work

- Children complete resource page A.

Plenary

- Ask the children to demonstrate, using the base 10 apparatus, how they solved each problem.

- *Who can think of other ways to answer these questions? Who can do it differently?*

- Discuss different methods for solving and recording the addition of three-digit numbers.

Name: _____

Adding with base 10

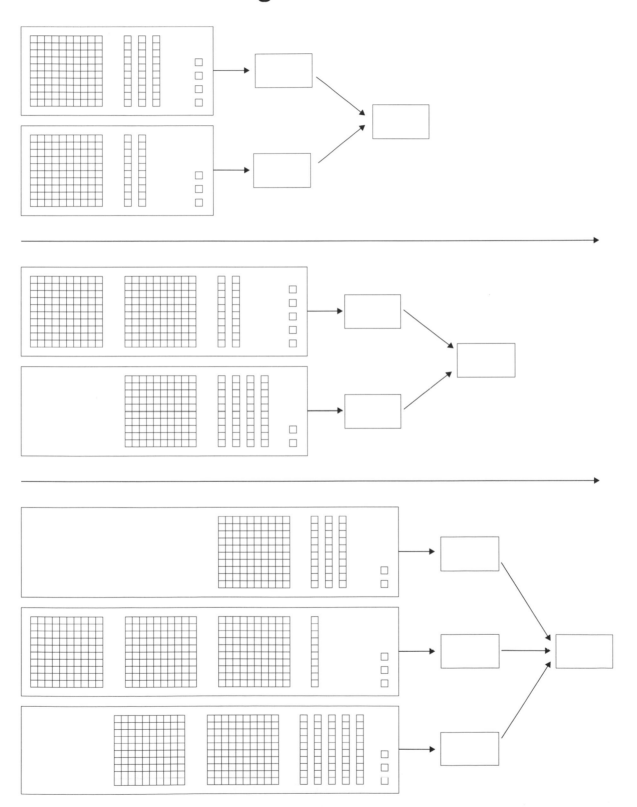

Adding two-digit numbers by partitioning into tens and ones

Advance Organiser

We are going to split two-digit numbers into tens and ones to find their sum

Oral/mental starter pp 184–185

Whole-class work

- Write on the board: *32.*

- *How can we write 32 as a number of tens and ones?*

- Record on the board: *32 = 30 + 2*. Do the same for 45.

- Question the children to check that they understand how to partition a two-digit number into tens and ones.

- Write on the board: *32 + 45*. *What is the answer to 32 add 45?*

- Discuss the different methods the children have used.

- Use their ideas to show them how partitioning works.

- Question the children as step-by-step you build up on the board:

- Explain that this 'vertical' way of recording shows how 32 + 45 is changed on each line to a new addition that has the same value as 32 + 45.

- Repeat with other additions of two-digit numbers, not crossing the 10.

$$32 + 45$$
$$= (30 + 2) + (40 + 5)$$
$$= (30 + 40) + (2 + 5)$$
$$= 70 + 7$$
$$= 77$$

Independent, paired or group work

- Ask the children to solve 41 + 26, 57 + 32, 25 + 63 and 84 + 15 in a similar way. You may wish to set out part of the working on the board.

- Extend to 36 + 23, 72 + 14, 63 + 31, 25 + 52, and so on.

Plenary

- Choose some examples to work through together.

- Continually ask the children to explain why they are using a particular stage of the calculation, how they know that that is what they should be doing, and so on.

Addition and Subtraction (5)

Outcome

Children will be able to use, record and explain mental and with-jottings methods of addition and subtraction

Medium-term plan objectives	• Extend understanding of addition and subtraction.
	• Add several small numbers.
	• Add or subtract a 'near multiple of 10' to a two-digit number, by adding or subtracting the nearest multiple of 10, and adjusting.
	• Use patterns of similar calculations.
	• Use informal pencil and paper methods to support, record and explain TU + TU, HTU + TU and HTU + HTU.
Overview	• Add 9, 19, 29, 39 and so on to a two-digit number by adding 10, 20, 30, 40 and so on and subtracting 1.
	• Subtract 9 from a two-digit number by subtracting 10 and adding 1.
	• Explain addition and subtraction patterns.
	• Extend addition and subtraction patterns.
	• Use more than one method to find TU + TU with carrying.

How you could plan this unit

	Stage 1	Stage 2	Stage 3	Stage 4	Stage 5
Content and vocabulary	Using a 1 to 100 grid to add near-multiples of 10 *addition, near-multiple of 10, move down and back*	Using a 1 to 100 grid to subtract near-multiples of 10 *subtraction, minus, near-multiple of 10, move up and on*	Finding answers in addition and subtraction patterns *addition patterns, subtraction patterns, extending patterns, explaining patterns*	Using different methods to find the answer to additions *addition, open number line, 1 to 100 number square, informal methods, apparatus*	
Notes					

Using a 1 to 100 grid to add near-multiples of 10

Oral/mental starter pp 184–185

Advance Organiser

We are going to use a 1 to 100 grid to help us do some additions

You will need: 1 to 100 number grid, paper arrows

Whole-class work

- Show the children a 1 to 100 number grid.

- Place an arrow starting at 43 and ending at 53.

- *What number does the arrow start at?*

- *What number does the arrow end at?*

- *How many added to 43 makes 53?*

- Explain that the arrow shows that 43 + 10 = 53.

- Leave the arrow in position for use in the next section.

- Repeat for other additions of 10. Each time, ask the children for the addition that the arrow shows.

- Repeat the activity with arrows for 2, 3, 4 and so on to show 'add 20, 30, 40' and so on.

1	2	3	4	5	6	7	8	9	10
11	12	13	14	15	16	17	18	19	20
21	22	23	24	25	26	27	28	29	30
31	32	33	34	35	36	37	38	39	40
41	42	43	44	45	46	47	48	49	50
51	52	53	54	55	56	57	58	59	60
61	62	63	64	65	66	67	68	69	70
71	72	73	74	75	76	77	78	79	80
81	82	83	84	85	86	87	88	89	90
91	92	93	94	95	96	97	98	99	100

- Record as 27 + 20 = 47, 65 + 30 = 95, and so on.

- Leave the arrows in position for use in the next section.

- With each addition, discuss that the tens digit increases by 1 and the unit digit stays the same.

- Repeat the activity but with two arrows, one to show 'add 10' followed by another to show 'minus 1' starting at 43.

- *What added to 43 gives 52?* Record as 43 + 10 – 1 = 52.

- Repeat the activity with arrows showing 'plus 19', 'plus 29' and so on, recording as 27 + 20 – 1 = 46, 65 + 30 – 1 = 94, and so on.

- With each addition, discuss that the tens digit increases by the number of tens added and the unit digit decreases by 1.

- Ask what causes each change.

Independent, paired or group work

- Children complete similar additions, such as 83 + 9, 49 + 19, 37 + 29, and so on.

- Allow them to use the 1 to 100 grids as support.

Plenary

- Ask mental additions, such as 23 + 49, 57 + 29, and so on. Record each addition and discuss what happens to the tens and to the unit digits.

Using a 1 to 100 grid to subtract near-multiples of 10

Advance Organiser

We are going to use a 1 to 100 grid to help us do some subtractions

Oral/mental starter pp 184–185

You will need: a 1 to 100 number grid, paper arrows

Whole-class work

- Show the children a 1 to 100 number grid. Place an arrow starting at 47 and ending at 37.

- *What number does the arrow start at? What number does the arrow end at?*

- *How many subtracted from 47 leaves 37?* Explain that the arrow shows 47 − 10 = 37.

- Leave the arrow in position for use in the next section.

- Repeat for other subtractions of 10. Each time ask the children for the subtraction that each arrow shows.

- Repeat the activity with arrows for 2, 3, 4, and so on to show 'subtract 20, 30, 40' and so on. Record as 78 − 20 = 58, 51 − 30 = 21, and so on.

- Leave the arrows in position for use in the next section. With each subtraction, discuss that the tens digit decreases by 1 and the unit digit stays the same.

- Ask the children to explain why. Repeat the activity but with two arrows, one to show 'subtract 10 followed by add 1' starting at 47.

- *What subtracted from 47 leaves 38?* Record as 47 − 10 + 1 = 38.

- Repeat the activity with arrows showing 'add 19', 'add 29', and so on, recording as 78 − 20 + 1 = 59, 51 − 30 + 1 = 22, and so on.

- With each subtraction, discuss that the tens digit decreases by the number of tens subtracted and the unit digit increases by 1.

- Ask the children to explain why.

1	2	3	4	5	6	7	8	9	10
11	12	13	14	15	16	17	18	19	20
21	22	23	24	25	26	27	28	29	30
31	32	33	34	35	36	37	38	39	40
41	42	43	44	45	46	47	48	49	50
51	52	53	54	55	56	57	58	59	60
61	62	63	64	65	66	67	68	69	70
71	72	73	74	75	76	77	78	79	80
81	82	83	84	85	86	87	88	89	90
91	92	93	94	95	96	97	98	99	100

Independent, paired or group work

- Children complete similar subtractions, such as 45 − 19, 37 − 29, 54 − 39, and so on.

- Allow them to use the 1 to 100 grids as support.

Plenary

- Ask mental subtractions, such as 63 − 49, 87 − 29, and so on. Record each subtraction and discuss what happens to the tens and to the unit digits.

Finding answers in addition and subtraction patterns

Advance Organiser

We are going to look at addition and subtraction patterns

Oral/mental starter pp 184–185

Whole-class work

- On the board, build up the pattern of additions shown.
- As each addition is written, ask for the answer.
- Discuss how the addition is the same or different to the one(s) that are above it.
- Draw the children's attention to how the digits change their place-value position.
- *What is the next addition in the pattern?*
- Record, and repeat the question.
- On the board, build up the pattern of subtractions shown.
- As each subtraction is written, ask for the answer.
- Discuss how the subtraction is the same or different to the one(s) that are above it.
- *What is the next subtraction in the pattern?*
- Record, and repeat the question.

$2 + 4 =$
$20 + 40 =$
$200 + 400 =$

$10 - 1 =$
$20 - 1 =$
$30 - 1 =$

Independent, paired or group work

- Ask the children to complete similar vertical sets of additions and subtractions.
- For example:

$5 + 4 = \square$	$8 - 3 = \square$	$19 - 8 = \square$
$65 + 4 = \square$	$28 - 3 = \square$	$29 - 8 = \square$
$365 + 4 = \square$	$528 - 3 = \square$	$39 - 8 = \square$
$2365 + 4 = \square$	$4528 - 3 = \square$	$49 - 8 = \square$

Plenary

- Together, go through the examples.
- Continually ask about changes to each place-value digit in the question and the answer.
- Try a slightly different example:

  $200 + 6 = \square$ $200 + 66 = \square$ $200 + 666 = \square$ $200 + 6666 = \square$

- Ask the children to explain how this is different or similar.

Using different methods to find the answer to additions

Advance Organiser

We are going to find the answers to additions using different methods

Oral/mental starter pp 184–185

You will need: apparatus, number lines or grids as required

Whole-class work

- Write on the board: *67 + 38*.

- Ask the children to work on their own to find the answer to 67 add 38 in any way they wish. Explain that they can use apparatus, number lines, a 1 to 100 number grid, diagrams, or make jottings if they wish.

- When they have finished, tell them to explain to a partner how they did it.

- Invite the children to explain how they did the addition.

- Encourage the other children to ask questions if there is anything they do not understand about the method.

- Keep a record of the different methods.

- Explain some of the methods that were described and others that the children may not have recalled. These may include the following examples of 67 and 55.

- Counting on in multiples on an open number line.

- Partitioning in tens and ones.

  $67 + 55 = (60 + 7) + (50 + 5) = (60 + 50) + (7 + 5) = 110 + 12 = 122$

- Partitioning into near-multiples of 5.

  $67 + 55 = 65 + 2 + 55 = 120 + 2 = 122$

- Identifying near-doubles.

  $67 + 55 = 60 + 7 + 60 - 5 = 120 + 2 = 122$

- Counting up to the next multiple of 10.

  $67 + 55 = 67 + 3 + 52 = 70 + 52 = 122$

Independent, paired or group work

- Ask the children to complete 84 + 67, 77 + 96, 58 + 85 and 66 + 77.

- They can use whatever method they prefer, but must record their method carefully.

Plenary

- Choose some of the examples. Compare the different methods the children have used.

- Explain to them that the best method for them is the one they can understand and use most successfully.

Addition and Subtraction (6)

Outcome

Children will be able to use, record and explain mental and supported methods of adding and subtracting

Medium-term plan objectives

- Add using pencil and paper methods.

- Use known facts and place-value to add/subtract mentally.

- Use informal pencil and paper methods to support, record or explain TU–TU and HTU–TU.

Overview

- Add a single-digit number to a three-digit number with no crossing of the tens boundary.

- Subtract a single-digit number from a three-digit number with no crossing of the tens boundary.

- Subtract, mentally, a single digit from a three-digit multiple of 100.

- Make a general statement about the unit digits and the tens digit of the answer when subtracting a single digit from a three-digit multiple of 100.

- Add 1/10/100 to a three-digit number.

- Make and test general statements about the unit/ten/hundred digit when adding 1/10/100.

- Subtract 1/10/100 from a three-digit number.

- Make and test general statements about the unit/ten/hundred digit when subtracting 1/10/100.

How you could plan this unit

	Stage 1	Stage 2	Stage 3	Stage 4	Stage 5
Content and vocabulary	Adding/subtracting a single digit to/from three-digit numbers *add, subtract, pattern, missing number*	Subtracting single digits from three-digit multiples of 100 *single digit, three-digit multiple of 100, general statement*	Adding 1/10/100 to three-digit numbers *general statement, change in the digit*	Subtracting 1/10/100 from three-digit numbers *general statement, change in the digit*	
Notes			Resource page A	Resource page B	

Add/subtract a single digit to/from three-digit numbers

Advance Organiser

We are going to add and subtract a single digit to and from a three-digit number

Oral/mental starter pp 184–185

Whole-class work

- Write on the board the first set of three additions shown.

- Ask the children to work out the answers for each addition.

- Invite the children to give the answers to each addition.

- *In what way are the three answers alike? Tell me why.*

- Discuss the different methods that the children used to solve the problems.

$$3 + 5 =$$
$$73 + 5 =$$
$$273 + 5 =$$

- Explain that in these calculations we are only adding a single-digit number, and the sum of the two unit digits in the two numbers is not more than 9. This means that any other digits, such as tens and hundreds, do not change.

- Repeat with the second set of additions, such as 7 + 2, 47 + 2 and 647 + 2.

- Now write on the board the first set of three subtractions shown.

- Ask the children to work out the answers for each subtraction.

- Invite the children to give the answers to each subtraction.

- *In what way are the three answers alike? Tell me why.*

- Discuss the different methods that the children used.

$$8 - 3 =$$
$$58 - 3 =$$
$$458 - 3 =$$

- Explain that we are subtracting a single digit from a number with a larger unit digit. Again, this means that any other digits, such as tens and hundreds, do not change.

- Repeat with the second set of subtractions, such as 7 – 5, 27 – 5 and 527 – 5.

Independent, paired or group work

- Ask the children to complete similar patterns of additions and subtractions.

Plenary

- Choose some examples to work through together.

- Remind the children that the tens and hundreds digits do not change. Discuss why this is.

Subtracting single digits from three-digit multiples of 100

Advance Organiser

We are going to subtract a single digit from a multiple of 100

Oral/mental starter
pp 184–185

Whole-class work

- Write on the board the first set of two subtractions shown.

- Ask the children to work out the answers for each subtraction.

- Invite the children to give the answers to each subtraction.

- Tell them to look for patterns in the two subtractions.

- *In what way are the two subtractions alike?*

- *In what way are the answers the same or different? Tell me why.*

- Discuss the different methods that the children used.

- Explain that we are subtracting a single digit from a multiple of ten or a hundred. This means that the multiple numbers have zero units and exchange is necessary.

- *How does the exchange take place in each subtraction?*

- *What can you tell me about the unit digits in the two three-digit numbers? (They add up to 10.) Why?*

- *What can you tell me about the tens digit when a single digit is subtracted from a three-digit multiple of 100?*

- Ask the children to use the pattern to work out the answer to 6000 – 7.

- Repeat with 40 – 4 and 400 – 4.

- Write on the board the general statement as shown:

- Invite the children to give you examples that test the truth of the general statement.

> $60 - 7 =$
> $600 - 7 =$

> When a single digit is subtracted from a three-digit multiple of 100 the sum of the single digit and the unit digit in the answer is 10

Independent, paired or group work

- Children complete similar sets of subtractions, such as 90 – 9, 900 – 9 and 9000 – 9; 600 – 3, 600 – 1, 600 – 6, and so on.

Plenary

- Choose some examples to work through together.

- Remind the children about the patterns in the subtractions.

Adding 1/10/100 to three-digit numbers

Advance Organiser

We are going to add 1, 10 and 100 to a number

Oral/mental starter pp 184–185

You will need: resource page A (one per child)

Whole-class work

- Write on the board the first set of three additions shown.
- Invite the children to give the answers to each addition.
- Discuss the different methods that the children used.
- *What changes when 1 is added to a number? Why?*
- *What changes when 10 is added to a number? Why?*
- *What changes when 100 is added to a number? Why?*
- Repeat with 618 + 1, 618 + 10 and 618 + 100.
- Write on the board the general statement: *When 1 is added to a number, the unit digit of the number is increased by 1.*
- Invite the children to give you examples to test the truth of this general statement.
- NOTE: Beware of additions such as 579 + 1. In such cases the general statement is not true, unless it is amended to 'numbers with a unit digit less than 9'.
- Ask the children to write general statements for adding 10 and 100 to a number.
- *When 10 is added to a number, the tens digit of the number is increased by 1.*
- *When 100 is added to a number, the hundreds digit of the number is increased by 1.*
- Test each general statement.
- Ask the children for numbers for which each general statement does not work.

> 345 + 1 =
> 345 + 10 =
> 345 + 100 =

Independent, paired or group work

- Ask the class to complete resource page A.

Plenary

- Choose some examples from resource page A. Invite the children to explain how they did the example.
- Question the children about the changes that take place in the unit/ten/hundred digits.

(**PUPIL PAGE**)

Name: _____

Adding 1, 10 and 100

Find the answers.

456 + 1 = ☐   826 + 1 = ☐   380 + 1 = ☐

456 + 10 = ☐   826 + 10 = ☐   380 + 10 = ☐

456 + 100 = ☐   826 + 100 = ☐   380 + 100 = ☐

Complete the missing numbers.

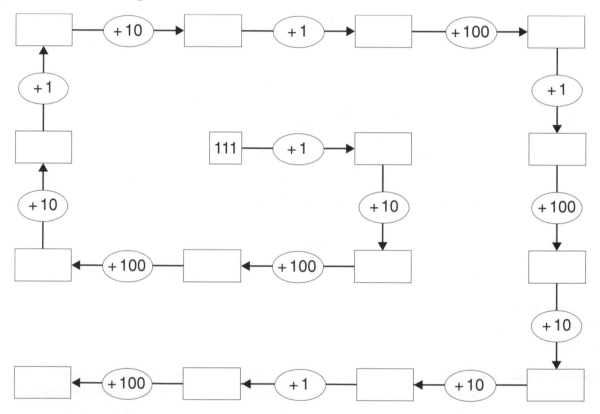

Find the answers.

409 + 1 = ☐   339 + 1 = ☐   769 + 1 = ☐

490 + 10 = ☐   393 + 10 = ☐   796 + 10 = ☐

4909 + 100 = ☐   3933 + 100 = ☐   7969 + 100 = ☐

Subtracting 1/10/100 from three-digit numbers

Advance Organiser

We are going to subtract 1, 10 and 100 from a number

You will need: resource page B (one per child)

Whole-class work

- Write on the board the first set of three subtractions as shown.

  526 – 1 =
  526 – 10 =
  526 – 100 =

- Invite the children to give the answers to each subtraction.

- Discuss the different methods that the children used.

- *What changes when 1 is subtracted from a number? Why?*

- *What changes when 10 is subtracted from a number? Why?*

- *What changes when 100 is subtracted from a number? Why?*

- Repeat with 384 – 1, 384 – 10 and 384 – 100.

- Write on the board the general statement: *When 1 is subtracted from a number, the unit digit of the number is decreased by 1.*

- Invite the children to give you examples to test the truth of the general statement.

- NOTE: Beware of subtractions such as 570 – 1. In such cases the general statement is not true, unless it is amended to 'numbers with a unit digit more than 0'.

- Ask the children to write general statements for subtracting 10 and 100 from a three-digit number.

- *When 10 is subtracted from a number, the tens digit of the number is decreased by 1.*

- *When 100 is subtracted from a number, the hundreds digit of the number is decreased by 1.*

- Test each general statement.

- Ask the children for numbers for which each general statement is not true.

Independent, paired or group work

- Ask the children to complete resource page B.

Plenary

- Choose some examples from resource page B.

- Ask the children to explain how they did each one.

- Question the children about the changes that take place in the unit/ten/hundred digits.

Name: _____

Subtracting 1, 10 and 100

Find the answers.

729 – 1 = ☐   265 – 1 = ☐   941 – 1 = ☐

729 – 10 = ☐   265 – 10 = ☐   941 – 10 = ☐

729 – 100 = ☐   265 – 100 = ☐   941 – 100 = ☐

Complete the missing numbers.

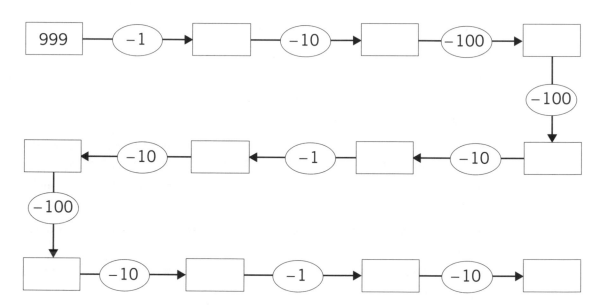

Find the answers.

660 - 1 = ☐   120 - 1 = ☐   790 - 1 = ☐

606 - 10 = ☐   102 - 10 = ☐   709 - 10 = ☐

6066 - 100 = ☐   1022 - 100 = ☐   7079 - 100 = ☐

Multiplication and Division (1)

Outcome

Children will be able to solve and represent multiplication and division using a range of visual models

Medium-term plan objectives

- Understand multiplication as repeated addition and as an array.
- Read and begin to write related vocabulary.
- Recognise that multiplication can be done in any order.
- To multiply by 10/100, shift the digits one/two places to the left.

Overview

- Relate multiplication to (repeated) addition and counting on.
- Investigate numbers as multiplication arrays.
- Use arrays to calculate multiplication and record the number sentences illustrating the commutative property.
- Reinforce understanding using other methods to check multiplications.

How you could plan this unit

	Stage 1	Stage 2	Stage 3	Stage 4	Stage 5
Content and vocabulary	Multiplication as repeated addition	Changing the order of calculation in multiplication	Multiplying and dividing by 10 or 100		
	lots of, groups of, times, multiply, multiplied by, repeated addition, array, row, column		*units, ones, tens, hundreds, digit, place, place value, one-, two- or three-digit number*		
Notes					

Multiplication as repeated addition

Advance Organiser

We are going to make rectangular arrays and number sentences

You will need: 4 × 6 array, squared paper, overhead projector (optional)

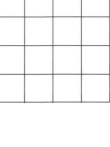

Whole-class work

- Show the children a 4 × 6 rectangular array, either on the board or shown on an OHT, and then turn it around to demonstrate the commutative property of multiplication.

- *My number is 24. What is my number now? Has it changed? How do you know?*

- *How could we write a number sentence to match this array? What about this way around?*

- Discuss with the class various number sentences for 24 and write them on the board. Include:
  4 + 4 + 4 + 4 + 4 + 4 = 24, 6 groups of 4 makes 24,
  6 × 4 = 24 and also 4 × 6 = 24.

- Discuss the difference between the last two number sentences. Ask the children to read them out in words, as 6 *lots of* 4 or 4 *groups* of 6. Relate to the rows and columns on the grid and confirm that the answer is the same each way.

- *Is there any other way you can group 24?*

- Rehearse counting in twos, threes, fours, and so on, asking the children to predict if they will say 24.

- Ask the children to investigate building rows of 2, 3 or 4 cubes or squares to make 24.

- *Write a number sentence for your rectangular array.*

- *How many multiplication number sentences can you write for 24?*

- Write the children's suggestions on the board and build up a list.

Independent, paired or group work

- Write the following numbers on the board: *15, 25, 36, 28, 45.*

- *Draw a rectangular array on squared paper for each number.*

- *Write some number sentences about your arrays.*

- Early finishers can choose prime numbers to investigate, or build squares of cubes and write other number sentences for those numbers.

Plenary

- Draw a rectangular array on the board.

- *How many squares are there? How did you know? Tell me a number sentence for this array.*

- *Tell me a different rectangular array for the same number.*

- *What are the number sentences for that array?*

Changing the order of calculation in multiplication

Advance Organiser

We are going to show that multiplication can be done in any order

Oral/mental starter p 185

You will need: dotted or squared paper

Whole-class work

- Use dotted or squared paper to draw a 3 × 2 array.

- *Adrian, come to the front of the class and write a number sentence for this array. What if the array was turned this way? What number sentence could we write then?*

- Confirm that you can write 3 × 2 = 6 and 2 × 3 = 6.

- *How can we check this is true?*

- Show the children, or draw on the board, a 0 to 20 number line.

- *On the line I am going to mark 3 jumps of 2.*

- *What number have I landed on?*

- *How else can I reach number 6 in jumps of equal size?*

- Include discussions of 6 jumps of 1, or 1 jump of 6.

- Write the number sentences on the board in pairs as shown.

- *What do you notice?*

- Repeat with other numbers, if necessary.

> 3 x 2 = 6, 2 x 3 = 6
> 6 x 1 = 6, 1 x 6 = 6

Independent, paired or group work

- Ask the children to draw rectangular arrays, number lines or other mathematical illustrations, demonstrating the following number sentences:
3 × 4 = 12, 3 × 5 = 15, 4 × 2 = 8, 3 × 10 = 30, 5 × 4 = 20, 2 × 6 = 12.

- *Write two number sentences for each illustration.*

- Early finishers can draw diagrams for square numbers.

- *What do you notice about the arrays?*

Plenary

- *If I write 2 × 5 = 10, who can tell me another multiplication sentence to which the answer is the same?*

- *Who can show one of these sentences as an array or on a number line?*

- *Can anyone show me another array or number line for the other sentence?*

- *If we know that 3 multiplied by 6 equals 18, what do we know about 6 multiplied by 3?*

- *How can we make sure we are right?*

Multiplying and dividing by 10 or 100

Advance Organiser

We are going to multiply and divide by 10 and 100 using a place-value chart

Oral/mental starter p 185

You will need: place-value chart, overhead calculator (optional), digit cards with extra zeros

Whole-class work

- Look at a place-value chart with the children, asking them what they notice about the position of the digits when a number is first multiplied by 10, and then by 100.

- Confirm that the digits move to the left. Use 0 as place holders for each 10 the original number is multiplied by.

- Continue to work with the chart asking the children to explain what happens when the number is divided by 10 then 100.

- Reinforce using an overhead calculator, if available.

- Play a game when the children are fairly confident about the movement and the direction of the digits when multiplied or divided by powers of 10.

- Give the children digit cards from 0 to 9 between them, with extra 0 cards.

- Call out a number and ask the children to make the number with their cards.

- Ask pairs to show their numbers to the class and check that everyone agrees.

- Call out a multiplication or division by 10 or 100; for example, *multiply by 100*.

- Using the place-value chart as a reference, the children agree on the new number and hold it up to show you. Go through various answers ensuring the class agrees on the correct answer and on how it was reached.

Independent, paired or group work

- Using place-value grids, the children record, by drawing arrows and filling in the boxes, the following calculations (one in each grid):
24 × 10, 320 ÷ 10, 560 ÷ 10, 23 × 10, 18 × 10, 450 ÷ 10.

Thousands	Hundreds	Tens	Units	
		2	4	x 10
	2	4	0	

Plenary

- *Can anyone tell us the answer to one of their calculations?*

- *Can anyone think of a word problem using those numbers?*

- *What is the answer to 240 × 100? Show me on the place-value chart what will happen. What about 24 ÷ 10?*

Multiplication and Division (2)

Outcome

Children will be able to understand multiplication and division as the inverse of each other

Medium-term plan objectives	• Understand division as grouping or sharing.
	• Read and begin to write the related vocabulary.
	• Recognise division as the inverse of multiplication.
	• Use doubling and halving, starting from known facts.
	• Say or write a division statement corresponding to a multiplication statement.
	• Choose appropriate number operations and calculation methods to solve problems.
Overview	• Division as grouping or sharing using, counting back on a number line.
	• Division as the inverse of multiplication, using counting back on a number line.
	• Double and halve, as strategies to simplify multiplications.
	• Solve problems using multiplication and division.

How you could plan this unit

	Stage 1	Stage 2	Stage 3	Stage 4	Stage 5
Content and vocabulary	Division as grouping or sharing *lots of, groups of, times, multiply, multiplied by, repeated addition, array, row, column, share, share equally, group in pairs, threes (and so on), equal groups of, divide, divided by, divided into*	Division as the inverse of multiplication *inverse*	Doubling and halving *double, halve*	Solving problems *what could we try next?, how did you work it out?, method, jotting, answer, show your working*	
Notes			Resource page A		

71

Division as grouping or sharing

Oral/mental starter
p 185

Advance Organiser

We are going to find out how many groups of 5 in 15

You will need: an empty number line

Whole-class work

- Using an empty number line, model division as repeated subtraction.

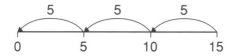

- *How many fives in fifteen?*

- Encourage the children to count on their fingers each time they count a jump of 5 backwards along the line. Check counting forward from 0.

- *There are three fives in fifteen.*

- Write on the board: *15 ÷ 3 = 5 and 15 divided by 5 equals 3 and there are 3 groups of 5 in 15.*

- Repeat using; *How many fours in twenty-four?* and so on.

- Demonstrate, asking the children to help you, counting backwards and forwards in fours from 24 along the line.

- Repeat using other divisions, ensuring that the children are counting the number of jumps on the line.

Independent, paired or group work

- Give the children an empty number line. Ask them to use the line, either by drawing jumps or working mentally, to solve:
  50 ÷ 10, 35 ÷ 5, 20 ÷ 4, 16 ÷ 4, 18 ÷ 3 and 21 ÷ 3.

Plenary

- Ask the/some children to come to the front of the class to demonstrate how they solved some of the problems.

- *Who can tell me a different way? Did anyone get a different answer?*

- Work through some examples with the children, demonstrating counting how many jumps it takes to get back to 0.

Division as the inverse of multiplication

Oral/mental starter p 185

Advance Organiser

We are going to find the inverse of 30 divided by 5

You will need: an empty number line

Whole-class work

- Demonstrate, using an empty number line, how to count backwards from 30 in groups of 5, asking the question; *How many groups of 5 are there in 30?*

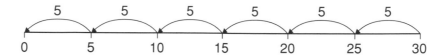

- Write the number sentence: *30 ÷ 5 = 6*, saying *30 divided by 5 equals 6*.

- The class then counts forward on the same number line in groups of 5 to show that 6 groups of 5 equals 30.

- Write the following number sentence on the board under the division: *6 × 5 = 30*.

- Continue to work with the class, counting forwards and backwards in groups of 2, 3, 4, 5 and 10, and recording the division calculation as well as the multiplication each time.

- Introduce the word 'inverse' to describe what is happening.

- *If we count forward 6 groups of 5, we reach 30. 6 times 5 makes 30. If we count back 6 groups of 5 starting at 30, we reach 0. 30 divided by 5 makes 6. This is the inverse calculation. Division is the inverse of multiplication. Multiplication is the inverse of division.*

Independent, paired or group work

- Lower-attaining children can be given number lines to perform the calculation whereas higher-attaining children can draw their own empty number lines.

- *Show how you would calculate the following divisions using a number line: 40 ÷ 10 =; 35 ÷ 5 =; 18 ÷ 2 =; 24 ÷ 4 =; 15 ÷ 3 =; 16 ÷ 4 =; 50 ÷ 5 =; 24 ÷ 3 =; 45 ÷ 5 =; 36 ÷ 4 =.*

- *Write the inverse number sentence each time.*

Plenary

- Ask the children to look at the following group of numbers and choose three to use together in a number sentence with the division sign: 2, 10, 4, 32, 16, 3, 100, 45, 5, 15, 20, 14.

- *Can you write a multiplication number sentence using the same numbers?*

- Ask the children to demonstrate how they worked out some of the inverses using a number line.

Doubling and halving

Oral/mental starter
p 185

Advance Organiser

We are going to use doubling and halving to make multiplying easier

You will need: resource page A (enlarged)

Whole-class work

- Draw a number line from 0 to 20 and ask the class to count in twos to 20. Highlight the multiples of two.

- Underneath, draw another number line to 20 and ask the children to help you highlight the numbers in the 4 times-table on it.

- *What is 1 lot of 4? What are 3 lots of 4? How can we find 5 groups of 4?*

- When the two lines are completed, ask the children what they notice.

- Draw out that doubling the 2 times-table will give the 4 times-table; multiplying by 4 is easy if you double then double again; halving the 4 times-table will give the 2 times-table.

- Show the children an enlarged copy of resource page A. Ask them what each part of the diagram shows.

- Try several numbers on the diagram to demonstrate.

- *How can what we know help us to multiply by 6? What can you tell me about the number 6?*

- Discuss with the class how multiplying by 6 could be made easier.

- Draw out that because 6 is double 3 and 3 is half of 6, we can multiply by 6 by doubling the 3 times-table.

- *What is 6 × 4? How could we use doubles or halves to help us?*

- Draw out and record that because 3 is half of 6, we can work out $3 \times 4 = 12$ and then double the answer to make 24.

- Check, using the 6 times-table.

Independent, paired or group work

- Ask the children to make the following multiplications easier to answer by halving so that a ×6 question becomes a ×3 question, and they double the answer: $8 \times 4 =$; $6 \times 6 =$; $5 \times 6 =$; $7 \times 4 =$; $9 \times 4 =$.

Plenary

- Ask the children to demonstrate how they worked out some of the questions.

- Draw diagrams similar to the one on resource page A to illustrate some of these relationships.

- *I know that $3 \times 5 = 15$, so I know that $6 \times 5 = 30$.*

- Write other multiplications on the board and ask the children what other facts they can derive from each.

Doubling and halving

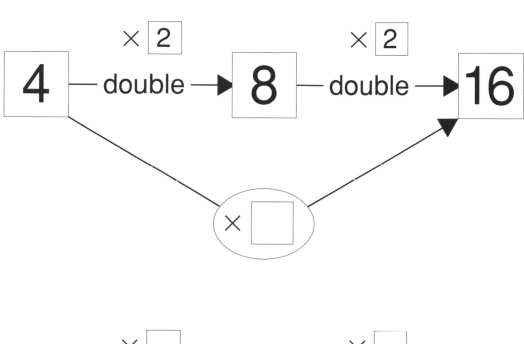

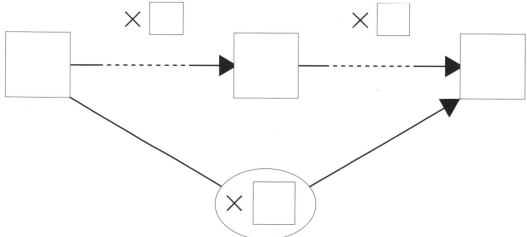

Solving problems

Advance Organiser

We are going to solve word problems by picturing in our heads what is happening

Oral/mental starter
p 185

You will need: counters

Whole-class work

- Read out the word problem below and ask the children to listen very carefully as you do so.

- *At 9 o'clock this morning I counted 22 bicycle wheels in the bike shed. How many bicycles were there?*

- Ask them to close their eyes and listen to you read the problem again. They should picture in their heads what is happening: are things being put together? Are things being split or shared? Are things being built up in groups or lots or sets?

- Ask the children to describe the problem as they have pictured it.

- *Dani, you told us that you pictured the wheels being joined together in twos. How could we model that picture to help us solve the problem?*

- Take other 'mental pictures' and model them using cubes or counters; for example, starting with 22 counters and counting them out in twos, then finding how many groups.

- *How could we write our problem as a number sentence?*

- Write on the board: $22 \div 2 = 11$, $11 \times 2 = 22$. There are 22 wheels and 2 wheels on every bike. So there are 11 bikes in the shed.

- Repeat for the following problem.

- *Aliens from Planet Pog have three fingers on each hand and five hands. How many fingers does one of these aliens have altogether?*

- *There are 16 boys in the class and four tables. An equal number of boys must sit at each table. How many boys sit at each table?*

- *There are five segments in each orange. I have four oranges. How many segments are there altogether?*

- Discuss each problem with the children, asking them to make a mental picture each time. They can record how their picture looks using words or jottings. Encourage them to look out for parts of the problem that are not relevant and to relate what is happening in the problem to an operation. Watch out for children thinking that 'altogether' always means 'add', and similar misconceptions.

Independent, paired or group work

- Ask the children to solve the word problems they have discussed above, in their pairs.

Plenary

- Work through the problems with the children. Add new features to the problems and discuss how to solve them; for example, *If two more Pog aliens join the group and each brings a dog how many Pog fingers are there now?*

Multiplication and Division (3)

Outcome

Children will be able to find remainders in context, and by using objects, and apply strategies and known facts to mental calculations

Medium-term plan objectives

- Begin to find remainders after division.
- Round up or down after division.
- Use known facts and place-value to multiply and divide mentally.

Overview

- Introduce remainders by sharing objects.
- 'Round' in context.
- Use partitioning to multiply.
- Use place-value charts to multiply and divide.

How you could plan this unit

	Stage 1	Stage 2	Stage 3	Stage 4	Stage 5
Content and vocabulary	Finding remainders *left, left over, remainder, divide, divided by, equal groups of*	Remainders in context	Calculation strategies for multiplication	Using place-value to multiply and divide mentally	
Notes					

Finding remainders

Advance Organiser

We are going to find how many are left over after a division

You will need: counters

Whole-class work

- Count out 29 counters with the children.

- *I want to divide these 29 counters into groups of 3. How many groups will that make?*

- Ask the children for their estimate or good guess.

- Ask a child to the front of the class to demonstrate grouping the counters into threes and taking a group away at a time. Stop when there are 9 groups of 3 and 2 counters left over.

- *What has happened? What should I do now?*

- Discuss that there are 9 groups of 3 counters, but also 2 counters left over. So there are nearly 10 groups of 3 counters.

- Write on the board: *29 ÷ 3 = 9 and 2 left over.*

- Repeat for grouping 26 counters in fours, again asking for a good guess or estimate first.

- *Joni, you estimate there will be more than 6 groups because you know that 6 groups of 4 makes 24. Who else has a good guess?*

- Demonstrate the grouping and write on the board: *26 ÷ 4 = 6 and 2 left over.*

- Repeat with other division calculations, asking the children to estimate first and to give reasons for their estimation.

Independent, paired or group work

- Ask the children to discuss a reasonable estimate to the following calculations, then calculate, using counters to find the remainders.

- Ask them to record their answers each time.

> $46 ÷ 5 =$, $27 ÷ 4 =$, $19 ÷ 3 =$, $32 ÷ 5 =$,
> $42 ÷ 4 =$, $54 ÷ 10 =$, $39 ÷ 1 =$

Plenary

- Ask the children to show you how they worked out some of their calculations.

- Write the answers on the board in the form *46 ÷ 5 = 9 and 1 left over*, and introduce the form *= 9 remainder 1*.

- Ask the children how they could check their work.

- Demonstrate checking, by multiplying and counting on; for example, checking that $27 ÷ 4 = 6$ remainder 3 by calculating $6 × 4 = 24$ and adding 3.

Remainders in context

Advance Organiser

We are going to decide how to round our answer to a division problem

You will need: counters

Whole-class work

- *There are 15 children at a party. At teatime, tables are set for groups of 4. How many tables do I need to seat all the children?*

- Ask the children to discuss the problem. They should close their eyes and picture what is happening as you read the problem.

- *Tania, how did you picture what was happening? Can you tell us what you think is happening?*

- Encourage the children to imagine the 15 children grouping in fours to sit at tables. Demonstrate with counters, or a diagram on the board if necessary.

- *So what happens after three tables are full? What happens to the three children left over? How many tables do they need?*

- Agree that although there are only three children left, they still need a whole table.

- *We need four tables to seat all the children. We round our answer up to give a sensible answer to the problem.*

- *I have 15 sweets. I want to give the same amount of sweets to each of my four children. How many sweets will they each get?*

- Again, encourage the children to visualise the problem. Demonstrate with counters, if necessary.

- *I share out the sweets one by one. After each child has three sweets, I have only three sweets left. Can each of the four children have one more sweet? No. So each child gets three sweets. We round our answer down to give a sensible answer to the problem.*

Independent, paired or group work

- Ask the children to work in groups to discuss, model and solve some of the following problems. They can act out the scenes, use counters, make diagrams, and so on.

- *Cakes are placed in boxes of 5. How many boxes do I need to hold 19 cakes?*

- *There are 15 boys in class 4. They are going to camp out in tents. If the tents are big enough for eight people, how many tents will we need for all the boys?*

- *I want to share 15 pies between two people equally. How many pies can they have each so that they have the same number?*

Plenary

- Ask some children to explain how they worked out the problems by acting out the scenes.

- The rest of the class can then suggest number sentences to describe the problem in its context.

Calculation strategies for multiplication

Oral/mental starter
p 185

Advance Organiser

We are going to practise some strategies for solving mental multiplications

You will need: arrow cards (place-value cards)

Whole-class work

- Give out arrow place-value cards and ask the children to use them to show you how various two-digit numbers are made up of tens and units.

- Encourage them towards the idea that, for example, 23 is made up of 20 and 3.

- *So what is the answer to 23 × 2? How could we solve that problem?*

- Encourage ideas from the children. If necessary, ask a pair of children to combine their place-value cards for the number 23.

- *How many 20-cards do you have? How many 3-cards do you have? What multiplication sentences could we write to help us?*

- Write on the board: *23 × 2 = 2 lots of 20 and 2 lots of 3.*

- Underneath, build up a partitioned multiplication to demonstrate the method as shown.

  23 × 2 = 2 × 20 and 2 × 3;
  20 × 2 = 40; 23 × 2 = 46.

- *How can we check we are right?*

- Talk through checking by sharing or grouping 23 by 2.

- Repeat for other two-digit by single-digit multiplications, as appropriate.

Independent, paired or group work

- Ask the children to work mentally, using jottings or place-value cards as appropriate, to write the answers to:
  14 × 2, 22 × 2, 31 × 5, 33 × 3, 12 × 4, 23 × 4.

Plenary

- Work through some of the children's answers with the class, pausing at every stage to ensure the children understand that because 23 is made up of 20 + 3, we can multiply 23 × 4 by adding together the answers to 20 × 4 and 3 × 4.

- Work through more examples, as appropriate.

Using place-value to multiply and divide mentally

Advance Organiser

We are going to use what we know about place-value to help us multiply and divide

You will need: place-value chart

Whole-class work

- Rehearse, with the class, multiplying and dividing by 10. Refer to a place-value chart, if necessary.

- *We know that 9×10 is 90. What is 9×5? How could we work it out?*

- Take suggestions from the children.

- Draw them towards the idea that because we know that 5 is half of 10, we know that 9×5 will be half of 9×10.

- *What about 90 divided by 10? The answer is 9. So what is the answer to 90 divided by 5?*

- Allow the children to discuss the problem.

- Lead the children towards the conclusion that because 5 is half of 10, we know that 90 divided by 5 will be double 90 divided by 10.

- Complete and compare number lines of the 5 times- and 10 times-tables with the class to show that the answer for dividing by 10 is half the answer when dividing by 5.

- *To divide by 5, divide by 10 and double the answer. To multiply by 5, multiply by 10 and halve the answer.*

Independent, paired or group work

- Write the following on the board and ask the children to write the answer to each number multiplied by 10 and multiplied by 5: *70, 30, 5, 40, 55, 23.*

- Write the following on the board and ask the children to write the answer to each number divided by 10 and divided by 5: *200, 250, 50, 750, 120.*

Plenary

- Ask the children to explain to the rest of the class how they worked out some of the problems.

- Work through some examples, reinforcing how each method works.

- *What is 130 divided by 5? How did you work that out? 130 divided by 10 makes 13, then double 13 makes 26. How can we check? How do we work out 26×5? Multiply 26 by 10, then halve the answer.*

- Encourage the children to check using another method, such as partitioning. For example, *26×5 is the same as 20×5 and 6×5, that makes $100 + 30$, which makes 130.*

Solving Problems (1)

Outcome

Children will be able to find totals, give change and apply different methods to solve money problems

Medium-term plan objectives	
	• Recognise all coins and notes.
	• Understand £p. notation (for example, £3.06).
	• Find totals, give change and work out how to pay.
	• Choose appropriate number operations and calculation methods to solve word problems. Explain and record methods informally.
	• Check sums by adding in a different order.

Overview	
	• Find totals.
	• Use £ signs to record money.
	• Work out possible totals from combinations of coins.
	• Solve word problems involving money.

How you could plan this unit

	Stage 1	Stage 2	Stage 3	Stage 4	Stage 5
Content and vocabulary	Finding and recording totals using coins and notes *note, value, amount*	Solving problems involving money *price, least/most expensive, price, costs more, costs less*	Solving word problems involving money		
Notes		Resource page A			

Finding and recording totals using coins and notes

Advance Organiser

We are going to make and record different totals

You will need: coins and notes, Blu-Tack or sticky tape

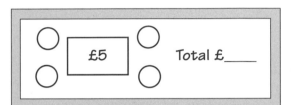

Whole-class work

- Draw a diagram on the board, as shown.

- *What can you tell me about this diagram?*

- Draw or stick different coins in the four circles around the note.

- *How much money is this? How did you work that out?*

- *How could we use the £5 note and four coins to make £6/£7?*

- *What's the smallest amount we can show with four coins and a £5 note?*

- *What is the largest amount we can make? What about the second largest? And the third? How can we work that out?*

- Work through with the children how to find the three largest amounts with a £5 note and any four coins.

- Encourage a systematic approach beginning with the largest four coins.

- *To find the second highest total we can make the highest total and then replace one coin with the coin of the next highest value; for example, replace a £2 coin with a £1 coin.*

- *Find the next highest total by swapping this £1 coin with a 50p coin.*

- Record the totals on the board.

Independent, paired or group work

- Ask the children to complete similar diagrams. For example, a £10 note in the centre, and £2 and 1p coins already marked; a £20 note in the centre, and a 5p coin already marked. Include a blank diagram.

- They should record how they made their totals each time.

- Early finishers could use the final diagram to set challenges for a partner; for example, they could place a £2 coin in one of the circles and ask their partner to make the smallest possible total.

Plenary

- Invite the children to give their answers and talk through how they worked out each set of totals.

- *Which coins did you use? Can you make the same total in a different way?*

- *How did you write that total? Did anyone have a different way of working?*

Solving problems involving money

Advance Organiser

We are going to find out, from the change we received, how much a book costs

You will need: money (coins and notes), Blu-Tack or sticky tape, resource page A

Whole-class work

- *I have £5. I buy a book and receive three coins as change. I am given a £2 coin and a £1 coin and one more coin. What could the book have cost?*

- Read through the problem and ask the children to close their eyes and visualise what is happening. Encourage them to think about what happens in the story.

- Draw the following diagram on the board and ask the children if they can explain what the diagram means in relation to the problem.

- Point to the £5 note.

- *I have £5.*

- Point to the middle square.

- *I buy a book.*

- *I was given three coins in my change.*

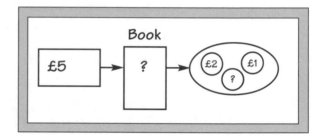

- Establish that you are told two of the coins, but that you need to list which the third coin might be. Draw a table on the board with four columns. In the first column write £2 + £1. In the second column list all the possible coins that could be added to this. Work through the totals with the children and list them in the third column.

- *Our first possible group of coins is £2 + £1 + £2. How much did that make? So how much did the book cost? Is this a sensible answer?*

- Move on to the second possible coin and repeat, recording how much the book must have cost and the appropriate number sentence in the final column; for example, £5 – cost of book = £4. *The missing number is £1, so the book cost £1.*

- Repeat for all the possible combinations of coins.

Independent, paired or group work

- Ask the children to complete resource page A.

- The final diagram is left blank and could be used by early finishers to explore the activity using a different note.

Plenary

- Invite the children to explain their answers.

- *How did you start? How do you know it can't be bought for less money?*

- Ask some early finishers to explain how they worked out their answers.

Name: _____

How much could each book cost?

Write the possible cost of each book in order starting with the least expensive.

This book could cost:

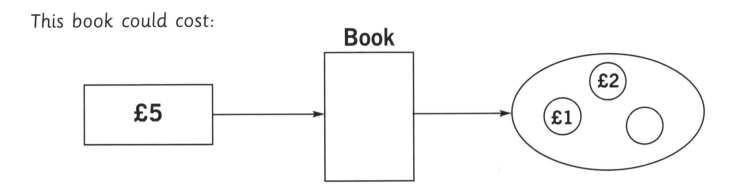

This book could cost:

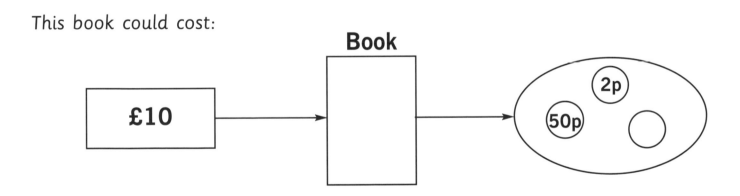

This book could cost:

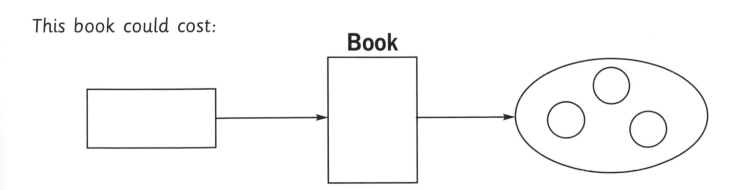

Solving word problems involving money

Advance Organiser

We are going to work out which books were bought

You will need: money (coins and notes), Blu-Tack or sticky tape

Whole-class work

- *I had £10 and bought two books. I got these coins in my change: £2, £1, 20p and 20p. Books cost £2.70, £3.50, £3.10 or £2.99. Which books did I buy?*

- Encourage the children to picture the problem while you read it.

- *Think about what is happening. I start with £10. I buy two books. I get back £2, £1 and two 20p coins. What do we need to find out first? How much were the books altogether?*

- Draw the following diagram on the board and draw or stick coins showing the change in the circle.

- *I had £10. I bought two books and was given this change.*

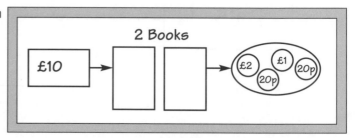

- Encourage the children to see how the diagram relates to the first stage of the problem. Work with them to add the coins to make £3.40, and then subtract £3.40 from £10 (or count on) to find out that the books cost £6.60 altogether.

- *What do we need to find out next? Books cost £2.70, £3.50, £3.10 or £2.99. Which two totals add up to £6.60 exactly?*

- Encourage the children to work systematically to find the totals of different combinations of books.

Independent, paired or group work

- Ask the children to use similar methods to solve other problems.

- *Malik bought 2 books. He was given £4.02 change. Which books did he buy? Maggie bought 2 books. She was given £3.80 change. Which books did she buy? Mabel bought 2 books. She was given £3 change. Which books did she buy? Michael bought 2 books. He was given £3.51 change. Which books did he buy? Melinda bought 2 books. She was given £4.31 change. Which books did she buy?*

Plenary

- Invite the children to explain how they worked out their answers.

- *How could you check that answer?*

- *If the change ends in 1p (for example, £3.51) the £2.99 book must have been bought. How do we know that?*

- *What do we know if the change given ended in 2p (for example, £4.02)?*

Solving Problems (2)

Outcome

Children will be able to record their work and explain how they reached and checked an answer

Medium-term plan objectives	
	• Choose an appropriate number operation and calculation method to solve word problems involving money and 'real life'.
	• Explain and record method informally.
	• Check multiplication in a different order.
	• Check subtraction with addition.

Overview	
	• Check multiplication calculations.
	• Check subtraction calculations.
	• Record working to explain methods.
	• Solve problems and check results.

How you could plan this unit

	Stage 1	Stage 2	Stage 3	Stage 4	Stage 5
Content and vocabulary	Recording working and checking results *calculate, calculation, mental, method, jotting, answer, right, correct, wrong, what could we try next? how did you work it out? number sentence, sign, operation, symbol*	Solving problems and explaining methods *tell me, describe, explain your method, show how you, show your working, record*			
Notes	Resource page A				

Recording working and checking results

Oral/mental starter pp 185–186

Advance Organiser

We are going to check our answers

You will need: resource page A (enlarged)

Whole-class work

- Show the children an enlarged copy of resource page A.

- *Explain what this diagram shows.*

- Discuss the first example with the children.

- *Who thinks this is the correct answer? Help me check this answer. How could we check to make sure?*

- Discuss various methods of checking. Write suggestions on the board. Point out that the working shows repeated addition of fours. Suggest that the children check, if they don't suggest it, using doubles of the 2 times-table, or repeated addition of sevens.

- *What is 2 × 7? And double 14 makes? So do we agree with the answer 29? Why not?*

- Discuss with the class what the working shows. Encourage them to spot what has gone wrong. Establish that the working begins on the number line at 1 and it should start at 0.

- Repeat for the lower half of the frame. Encourage the children to check the result by adding 9 and 26.

- Encourage children to see where the mistake has been made: the working shows taking away 10, then taking away 1 more. It should show taking away 10 then adding 1.

- Work through 8 × 3 and then 41 – 17 on the board with the children, recording the methods suggested.

- When you have reached an answer, ask the children to suggest how to check that the result is correct.

Independent, paired or group work

- Ask the children to record their working in similar ways to solve 9 × 5, 83 – 34, and 6 × 7.

Plenary

- Invite the/some children to show the class their working and to explain what they did.

- *Who can check this answer for me? Did anyone get a different answer? Does anyone know another way?*

(EXAMPLE)

Problem-solving frame

Number sentence $\qquad 4 \times 7 = \boxed{}$

Working

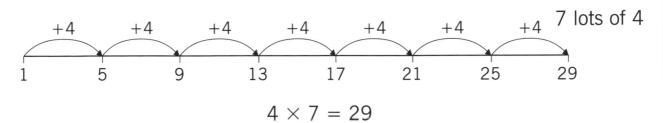

7 lots of 4

$4 \times 7 = 29$

Check

Number sentence $\qquad 37 - 9 = \boxed{}$

Working

$37 - 10 = 27$

$- 1$ more

$27 - 1 = 26$

$37 - 9 = 26$

Check

Classworks © Classworks Numeracy author team, Nelson Thornes Ltd, 2003

Solving problems and explaining methods

Advance Organiser

We are going to solve problems and record and explain our work

Whole-class work

- Read the following problem to the children. Encourage them to close their eyes and picture what is happening.

- *A teacher has 32 children in her class. Each child has brought 5p for a charity collection. How much money is there altogether?*

- Discuss with the children how they are picturing the problem.

- *How do you see the problem, Harry? Describe in your own words what we have to find out.*

- Encourage the children to discuss ways of solving the problem and to relate what happens to mathematical operations.

- Lead them towards the idea that they need to multiply the 5p each child brings by the number of children (32) in the class.

- Write this information on the board in a number sentence: $32 \times 5p = \boxed{}$.

- *How can we work this out?*

- Encourage the children to think of mental, or mental-with-jottings ways to work out $32 \times 5p$. Suggest a method they have used before, such as multiplying by 10 and halving, or multiplying by 30 and by 2 and adding the results.

- Record the method chosen on the board. *How can we check our result?*

- Encourage the children to think of a second way of solving the problem and record that as well. *Do we have the same answer? What should we try next?*

- Continue checking and recording working until all the children are happy with the result.

Independent, paired or group work

- Ask the children to answer the following word problem:
  There are 14 tables in the class. Four children can sit at each table. What is the largest number of children that could be seated at the tables?

- They should follow the methodology above and record their work twice: once to get one answer, once to check. Read out the problem to them first and ask them to visualise what happens.

Plenary

- Invite the children to explain their answers, and how they reached them.

- Ask the class for alternative methods, answers, checks and explanations.

- Ask two children to swap record sheets and explain each other's method to the class.

- *Was it easy to understand what Jimena had done? Why do you think that is? Who can check Sufi's work?*

- Encourage the children to relate the recording to the ease of explanation.

Solving Problems (3)

Outcome

Children will be able to solve multiple-step problems and record their work

Medium-term plan objectives	• Choose appropriate number operations and calculation methods to solve money or 'real-life' word problems with one or more steps.
	• Explain and record method.
	• Check with an equivalent calculation.
Overview	• Choose calculation methods.
	• Explain and record.
	• Solve word problems.
	• Use diagrams to record problem solving.

How you could plan this unit

	Stage 1	Stage 2	Stage 3	Stage 4	Stage 5
Content and vocabulary	Two-step problems *add, plus, sum, total, take away, take, subtract, minus, sign, multiply, times, double, halve, divide, divided by, share, shared by, equals*	Solving 'think of a number' problems *inverse*			
Notes					

Two-step problems

Oral/mental starter pp 185–186

Advance Organiser

We are going to solve a problem in two steps

Whole-class work

- *I have 17 stickers. Jasmine says she will give me 5p for every sticker I give her, and double the total amount if I give her a sticker album as well.*
  How much will I get if I give her 10 stickers and a sticker album?
  What if I give her 12 stickers and a sticker album?
  What is the most money I can get without giving her the sticker album?

- Write the above on the board and read the question out. Ask the children to picture, in their heads, what is happening.

- Discuss the problem with the children.

- *What do we need to do to solve the problem? Can we make it easier?*

- If necessary, suggest breaking the problem into stages.

- Draw a two-step number machine on the board as shown.

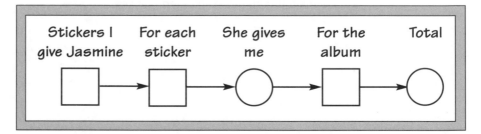

- Point to the first box.

- *This is the amount of stickers I could give Jasmine.*

- Point to the first part of the number machine.

- *What do I write here? How do you know?*

- Encourage the children to see that you need to write × 5p on this section because Jasmine gives you 5p for each sticker.

- *We write the answer to this step in this circle. What happens at the next stage?*

- Draw out that the next stage should be double whatever amount comes out of the first stage.

- Work through the first example with the class, working out 10 × 5p then doubling the answer.

- Ask for alternative ways of solving the problem – some children may see that a quicker way would be to work out 10 × 10p, which is the same as double 10 × 5p.

Independent, paired or group work

- Ask the children to solve the other two parts of the problem. Encourage them to use similar diagrams or other methods if they feel that is easier.

Plenary

- Invite the children to explain how they found their answers.

- Demonstrate checking their answers by working backwards (halving then dividing).

Solving 'think of a number' problems

Advance Organiser

We are going to work out the mystery number

Whole-class work

- *I want you to close your eyes and picture what is happening. I'm thinking of a number. I add 3 to it, then multiply by 2 to make 26. What number was I thinking of?*

- Discuss how to solve the problem with the children. Ask some of them to explain, in their own words, what they think the problem is.

- *How could we make it easier to solve the problem?*

- Draw the following diagram on the board and suggest that the children break the problem into two steps.

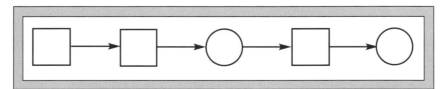

- Point to the first box.

- *I don't know what the first number is, so I leave this box blank.*

- *I add 3 next; I'll write + 3 in the next box. The next box is for the answer to this part of the question so I'll leave it blank for now.*

- *I then multiply my answer by 2; I'll write × 2 in this rectangle. What do I write for my answer in the last box?* Write 26 in the last box. *How do I solve the problem now?*

- Agree that the children should work backwards using the inverse operation each time. They halve or divide 26 by 2, and write the answer (13) in the middle box. They then subtract 3 to get to the original number.

Independent, paired or group work

- Ask the children to solve further 'think of a number' puzzles.

- For example, *I'm thinking of a number. I add 2 then multiply by 2 to make 26. What number was I thinking of?*

- Repeat for add 2 then multiply by 2 to make 54, add 7 then multiply by 2 to make 38, add 13 then multiply by 2 to make 88, add 22 then multiply by 2 to make 104.

Plenary

- Invite the children to show how they solved each problem.

- Work through one of the problems with the class.

Solving Problems (4)

Outcome

Children will be able to solve 'real-life' problems by applying a range of strategies

Medium-term plan objectives

- Choose appropriate number operations and calculation methods to solve money or 'real life' word problems with two steps.

- Explain and record method.

- Check results; for example, check division by multiplication, halving by doubling.

Overview

- Solve word problems using jottings.

- Record working and check answers.

- Use and apply multiplication and division to solve problems.

- Check results.

How you could plan this unit

	Stage 1	Stage 2	Stage 3	Stage 4	Stage 5
Content and vocabulary	Visual strategies for two-step problems *times, multiply, multiplied by, once, twice, three times (and so on), double, times-table, halve, share, share equally, divide, divided by*	Solving division problems			
Notes		Resource page A			

94

Visual strategies for two-step problems

Oral/mental starter
pp 185–186

Advance Organiser

We are going to solve a real-life problem

You will need: counters

Whole-class work

- Read the following word problem to the children. Ask them to close their eyes as you read it and picture what is happening as you do so.

- *Forty-two children go on a school trip. They have to stay in pairs the whole time. There is one minibus that seats 25 people, one that seats 12 people and one that seats nine people. There must be at least one space for an adult left over in each minibus. How many children can sit in each minibus in their pairs?*

- Encourage the children to discuss what they have to do to solve the problem and how they picture it. Note the key facts on the board.

- *What is happening in this problem? Are things being brought together at any point? Are things being shared out? How could we make it easier to solve? How many parts are there to the problem?*

- Lead children towards the idea that they need to find how many pairs of children can fit in each minibus to make a total of 42 and still have at least one space left over in each minibus.

- Draw the following diagram on the board to illustrate the problem.

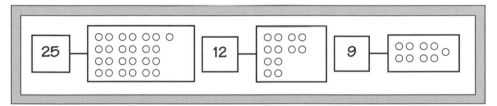

- *How does this help us? Is this how you visualised the problem? Did anyone have a different way of picturing what is happening?*

- Write and discuss further ideas. Encourage children to use pairs of counters and to draw their own 'minibuses' to model the problem.

- Help the children to recognise that in the context of the problem the minibus for 12 can only seat five pairs of children, because there must be a remainder of at least one space for an adult. The solution is that 24 children are in the first bus, 10 children in the second and eight in the third.

Independent, paired or group work

- Ask the children to solve similar problems, recording their working each time. For example: *There are 17 red apples and 26 green apples. How many children can have three apples each?*
  There are 47 cola bottles, 25 flying saucers and 19 toffees. How many party bags can I make with four sweets in each bag?

Plenary

- Invite some children to the front of the class to explain their answers and how they reached them. *How can we check our answers? Did anyone work differently?*

- Work through, checking some answers using the inverse operation.

Solving division problems

Advance Organiser

We are going to work out which number divided by 3 makes 5

You will need: counters, resource page A

Whole-class work

- *A number was shared by 3 to make 5. What was the number?*

- Discuss how to solve the problem with the children. Encourage them to visualise what is happening. Write it on the board as a 'missing number' question:

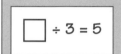

- *What divided by 3 makes 5?*

- Draw the children towards the idea that they can use the inverse of the operation to find the answer.

- *If my number divided by 3 makes 5, then 5 times 3 makes my number.*

- Write on the board underneath the first equation: $5 \times 3 = \square$.

- Draw two function machines to illustrate further what is happening.

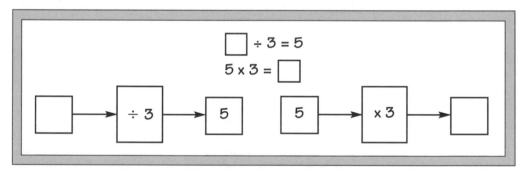

- *What does this diagram show? Who can tell me how this helps us?*

- Repeat for other missing-number division problems.

- *A number was shared by 2 to make 10. What was the number?*

Independent, paired or group work

- Ask the children to complete resource page A. Encourage them to make jottings and draw diagrams to support their work.

- Support the children by suggesting visual frames such as those shown above. Early finishers could write similar problems for a partner to solve.

Plenary

- Invite the children to explain how they checked their answers.

- Work through an example of checking with the children.

- *We worked out that $\square$ shared by 2 makes 6 by multiplying 6×2 to find 12. How can we check? What is 12 shared by 2? How else could we check?*

- Draw out suggestions such as halving 12, and dividing 12 by 6.

Name: _____

Solving division problems

Write the answer in the box each time. Check every answer and record how you checked it.

| | shared by 2 makes 6. |

| | shared by 2 makes 14. |

| | shared by 2 makes 19. |

| | shared by 2 makes 16. |

| | shared by 2 makes 17. |

| | shared by 2 makes 26. |

| | shared by 3 makes 9. |

| | shared by 3 makes 6. |

| | shared by 3 makes 11. |

| | shared by 3 makes 15. |

Solving Problems (5)

Outcome

Children will be able to identify and find the information needed to solve word problems

Medium-term plan objectives

- Choose appropriate number operations and calculation methods to solve money or 'real-life' word problems with one or two steps.

- Explain and record method.

- Check results.

Overview

- Use tables of answers.

- Check answers.

- Record methods.

- Identify the correct information to solve problems.

How you could plan this unit

	Stage 1	Stage 2	Stage 3	Stage 4	Stage 5
Content and vocabulary	Organising answers to problems *times, multiply, multiplied by, times-table, add, altogether*	Identifying what you need to find out, to solve a problem *what can we try next? what do we need to know?*			
Notes	Resource page A				

98

Organising answers to problems

We are going to organise our answers to solve problems quickly

Oral/mental starter pp 185–186

You will need: coins, Blu-Tack and paper, resource page A (one per child)

Whole-class work

- Write a price list for a swimming pool on the board, such as: *Child tickets: 30p, Adult tickets: 50p.*

- *Who can tell me what this shows?*

- Establish that the picture shows two types of ticket for a swimming pool.

- *How much would it cost to buy three child tickets? How did you work that out? How much would two adult and one child ticket cost altogether?*

- Ensure that the children are able to derive the correct information from the price list.

- *How could we make it easier to solve these questions quickly? Do we have to work them out each time?*

- Draw a simple table on the board.

- *How does this help us to find answers? Who can show me how to use it? How much would three child tickets and two adult tickets cost altogether?*

- *I spent £1.50 on three tickets. Which tickets could I have bought?*

- Discuss with the children how you can find amounts on the table and derive the tickets bought from that.

- *I spent £1.80 on four tickets. Which could I have bought?*

- Ensure the children check to make sure there is only one answer.

Number of tickets	Children 30p	Adults 50p
1	30p	50p
2	60p	£1
3	90p	£1.50
4		

Independent, paired or group work

- Ask the children to complete resource page A.

- Tell them to begin by drawing a table of answers similar to the one above.

Plenary

- Invite the children to show their tables and explain how they used them.

- *Did you need to make a table showing the price of 10, or 30 tickets? Why? Did anyone find a quicker way?*

- Show them that the maximum number of tickets referred to in the questions is seven.

- Ask some quick questions based on the tables.

(**PUPIL PAGE**)

Name: _____

Swimming pool problems

Write which tickets each family bought.

Swimming Pool Admission Prices	
Adults	£1.05
Children	60p

The Homer family paid £3.30 for 4 tickets.

They bought:

The Burns family paid £3.15 for 3 tickets.

They bought:

The Beaney family paid £2.85 for 4 tickets.

They bought:

The Smithers family paid £4.80 for 5 tickets.

They bought:

The McLoud family paid £3.45 for 5 tickets.

They bought:

The Drudge family paid £5.10 for 7 tickets.

They bought:

Classworks © Classworks Numeracy author team, Nelson Thornes Ltd, 2003

Identifying what you need to find out, to solve a problem

Advance Organiser

We are going to find out how many of each ticket we bought

Oral/mental starter pp 185–186

You will need: coins

Whole-class work

- *Entrance to the museum costs £1 for adults and 60p for children. How much will two adult tickets and one children's ticket cost altogether?*

- Ask the children to close their eyes and visualise the problem. Discuss with the children various ways of solving the problem. Write suggestions on the board.

- Suggest using a table to record the problem. Write the following table on the board.

- Discuss with the children how the table represents the problem. Ensure they can read across that the cost of each ticket is in the first column.

Adult £1	2	
Children 60p	1	

- *How do we find the total? Show me how you worked it out.* Complete the table.

- Change the information in the problem and ask the children to solve it using the table.

- *I spent £3 on adult tickets to the museum. I spent £4.20 altogether.*

Adult £1		£3.00
Children 60p		
		£4.20

- Show the children how to change the information on the table and work out what is missing.

Independent, paired or group work

- Ask the children to complete similar problems. They can use coins and a similar table to help them.

- *I bought four adult tickets. I spent £2.40 on children's tickets. How much did I spend altogether?*

- *I bought three children's tickets. I spent £5.80 altogether. How many adult tickets did I buy?*

- *I bought three adult tickets and six children's tickets. How much did I spend altogether?*

- *I spent £5.60 altogether. Which tickets did I buy?*

Plenary

- Invite the children to explain how they worked out their answers.

- Invite the children to show any answer tables they used.

- *Which part of the question did you work out first?*

Solving Problems (6)

Outcome

Children will be able to choose methods and record their work so they can explain it later

Medium-term plan objectives

- Choose appropriate number operations and calculation methods to solve money or 'real-life' word problems with one or two steps.
- Explain and record method.
- Check results.

Overview

- Check halving by doubling in context.
- Explain methods.
- Solve word problems involving multiplication.
- Apply visual strategies to word problems.

How you could plan this unit

	Stage 1	Stage 2	Stage 3	Stage 4	Stage 5
Content and vocabulary	Problems involving doubles and halves *grid, row, column, double, twice, two times, halve, half, shared by two*	Money problems involving multiplication *lots of, groups of, times, product, multiply, multiplied by*			
Notes					

Problems involving doubles and halves

Advance Organiser

We are going to find how many animals altogether

You will need: counters

Whole-class work

- *In the St Thomas Home for Elderly Badgers, each pair of badgers gets a pet squirrel to keep them company. If there are 22 badgers altogether, how many squirrels are there? How many animals altogether?*

- Encourage the children to discuss how to solve the problem. Take ideas and write them on the board. Encourage the children to visualise the problem with their eyes closed as you read it out.

- Discuss organising the information on a table or grid to help with the visualisation, or using counters. Draw a diagram on the board as shown.

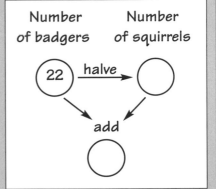

- *I counted the animals in the home and there were 12 altogether. How many badgers and how many squirrels is that? How can we work it out?*

- Rewrite the diagram with 12 in the final circle, and the others blank.

- Establish that to solve the problem the children have to find two numbers, one double the other, that add to make 12. Some children may be able to divide the total by 3 and use that as the smaller number in the problem. Explain that this is because 2 lots and 1 more lot of a number makes 3 lots altogether.

- Write on the board: *double 5 plus 5 is the same as 3 lots of 5.*

- Support the children by showing them how to use a set of counters to make the total. The counters need to be shared between two sets, one of which is twice as big as the other.

Independent, paired or group work

- Ask the children to solve similar word problems.

- *Every squirrel in the park collects two hazelnuts. If there are 30 hazelnuts altogether, how many squirrels are there?*

- *For every bottle of lemonade I buy, I also buy two bottles of mineral water. If I have 27 bottles altogether, how many bottles of each did I buy?*

- *My aunt gives me 2p for every 1p I have saved. Now I have 90p altogether. How much had I saved before my aunt gave me any money?*

Plenary

- Invite the children to explain how they solved the puzzles.

- Use counters to model solutions.

- *How could you check your answer?*

- Work through checking some of the solutions with the class.

Money problems involving multiplication

Oral/mental starter
pp 185–186

Advance Organiser

We are going to find how much money Tariq has to raise

You will need: counters

Whole-class work

- *Tariq is raising money to go on a sponsored walk. For every £1 sponsorship he raises himself, the government will give him £3. In order to go on the sponsored walk, he has to raise £100 altogether. How much does he have to raise himself? How much must the government give him?*

- Discuss with the children how to solve the problem. Ask them to visualise what is happening. Invite the children to explain how they would work it out.

- Draw the following diagram on the board.

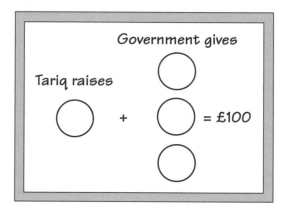

- *Did anyone visualise this happening? Who can tell me what this means?*

- Discuss how to solve the problem. Allow the children to use counters to model the problem.

Independent, paired or group work

- Ask the children to solve the problem.

- Encourage the children to see that there are four lots of £1 involved in every stage of the problem: £1 provided by Tariq plus £3 provided by the government. They can solve the problem by dividing £100 by £4 and using this as how much Tariq raises. They then multiply by three to find how much the government contributes.

Plenary

- Invite the children to explain how they solved the problem.

- *What did you do first? Why did you decide to do that?*

- *How could you check your answers?*

Measures (1)

Outcome

Children will be able to measure time to 5 minutes, to draw lines to within half a centimetre and to record lengths to the nearest half unit and as mixed units

Medium-term plan objectives

- Read time to 5 minutes.
- Use a ruler to draw and measure lines to the nearest half centimetre.
- Read and begin to write the vocabulary related to length.
- Choose an appropriate number operation and calculation method to solve word problems.
- Explain and record method informally.
- Measure and compare using metres and centimetres.
- Know the relationship between metres and centimetres, kilometres and metres.
- Use decimal notation for metres and centimetres.
- Suggest suitable units and equipment to estimate or measure lengths, including kilometres.
- Read scales.
- Record to the nearest whole/half unit, or as mixed units (e.g., 3m and 20cm).

Overview

- Read analogue clock times to 5 minutes.
- Use a ruler to draw lines of specified lengths.
- Record lengths to the nearest half unit and to mixed units.

How you could plan this unit

	Stage 1	Stage 2	Stage 3	Stage 4	Stage 5
Content and vocabulary	Time in words and numbers *hour, minute, second, o'clock, clock, watch, hands*	Measuring and drawing lines *length, width, height, depth, long, short, tall, high, low, wide, narrow, centimetre, ruler*	Measuring in metres and centimetres *metre, metre stick, tape measure*	Introducing decimal notation	Introducing the kilometre
Notes	Resource page A	Resource pages B, C and D		Resource page E	Discuss very long distances (to the next town, nearest country, moon) and introduce a kilometre as one thousand metres. Discuss things you would measure in kilometres.

Time in words and numbers

Oral/mental starter
p 186

Advance Organiser

We are going to write the time in words and in numbers

You will need: analogue clock face with geared hands and minutes numbered in fives, resource page A (one per child and one enlarged)

Whole-class work

- Place the hands of the analogue clock at 1 o'clock.
- *What time does this clock show?*
- Move the minute hand to 5 past.
- *Does anyone know what time the clock shows now?*
- Take ideas and agree that this is 5 minutes past 1.
- Write on the board: *five past one and 1.05.*
- Continue moving the hands of the analogue clock face around very slowly.
- As the minute hand passes each number, count an extra 5 minutes past.
- *Five past 1, 10 past 1, 15 past 1, and so on.*
- Remind the children that there are 60 minutes in an hour.
- Count round through one hour: 50 minutes past 1, 55 minutes past 1, 2 o'clock.
- Now look at the clock face with the minutes marked on.
- Count around in fives together pointing at each numeral as you do so.

Independent, paired or group work

- Ask the children to complete resource page A, recording the times in words and numbers.
- Let them look at the analogue clock face if necessary to help them count around in fives each time.

Plenary

- Go through a display version of resource page A with the children, inviting them to insert the missing times.
- Let one child call out a 'fives' time and another can set the clock to match, with the help of the class.

Name: _____

Minutes past

Write in the times using words and numbers as the example shows.

Twenty past three

_____ 3 : 20 _____

_____ : _____

_____ : _____

_____ : _____

_____ : _____

_____ : _____

_____ : _____

_____ : _____

_____ : _____

_____ : _____

_____ : _____

_____ : _____

Measuring and drawing lines

Oral/mental starter
p 186

Advance Organiser

We are going to measure and draw lines accurately

You will need: rulers, fine felt tip or ballpoint, resource page B (enlarged), resource page C (one per child), resource page D (one per child)

Whole-class work

- Give each child a ruler to look at while you talk about it.
- Draw the children's attention to the space at each end, the start marked with a zero followed by the numbers 1 to 30.
- If there are half centimetres marked, point these out.
- Remind the children that cm is used as a short form of centimetres.
- Show the children the enlarged version of resource page B.
- Make sure they all understand it has been magnified so that they can see it better.
- *How long is this line?*
- Reinforce, depending on answer(s), that it is halfway between 9 and 10 cm.
- *So how long must the line be?*
- Confirm the line is $9\frac{1}{2}$ centimetres long.
- Write on the board: $9\frac{1}{2}$ *cm*.
- Draw a few lines on display paper using a ruler and ballpoint or fine felt tip.
- Ask various children to measure them, checking that they are using the ruler correctly.

Independent, paired or group work

- Ask the children to measure the lines on resource page C and write the answers in the box provided.
- An extra activity is the jumping game on resource page D. The baby frogs are trying to jump onto a lilypad.
- Each frog has their own starting point. The children need to find out where each frog lands by drawing a line of the correct length towards the lilypad.
- Demonstrate how to align the ruler with zero on the frog's spot, towards the target, and draw a line of the correct length.

Plenary

- Take some of the children's answers and see if there are any differences.
- Demonstrate checking.
- *Who agrees? Does anyone else want to check?*
- *Can you describe how you measured?*
- Place an acetate version of resource page D over some children's sheets.
- *Which frog jumped the furthest/shortest distance?*

PUPIL PAGE

Name: _____

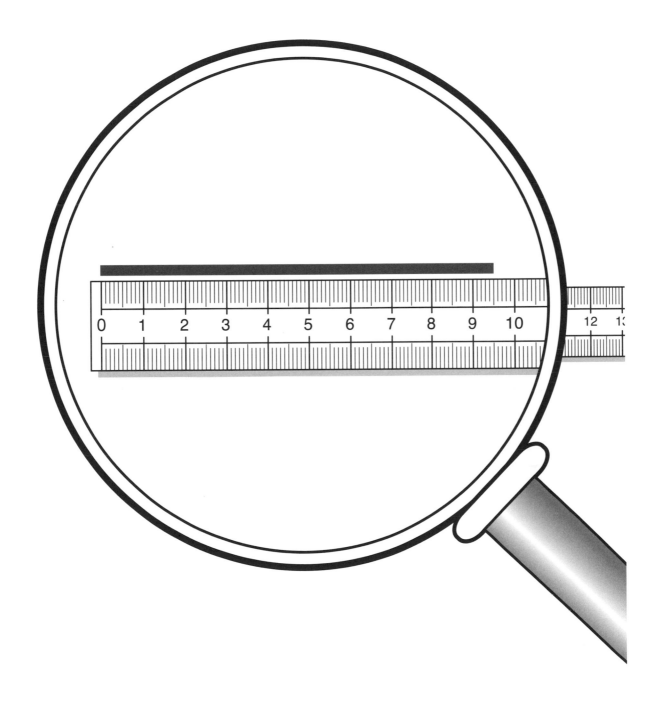

Classworks © Classworks Numeracy author team, Nelson Thornes Ltd, 2003

(**PUPIL PAGE**)

Name: _____

Measuring lines

Write the length of each line like this: [5] cm

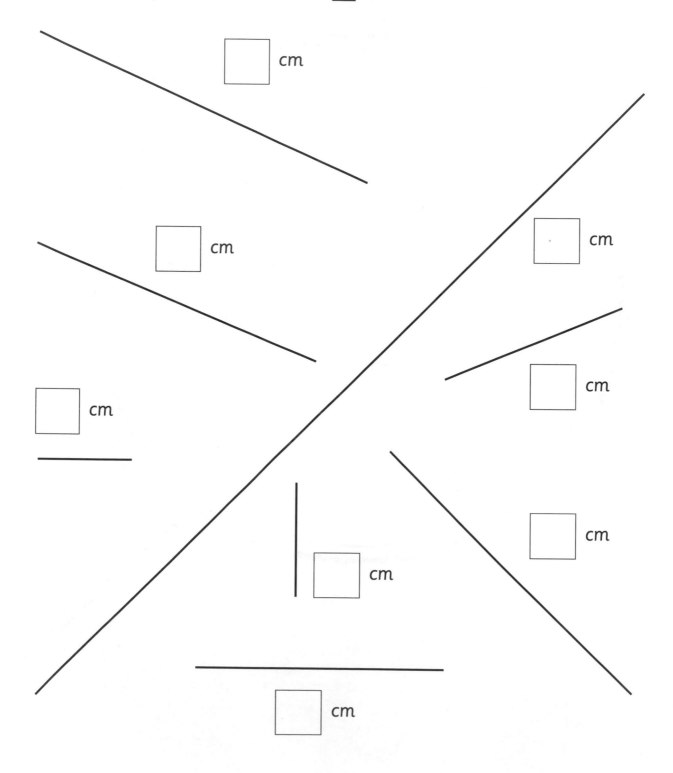

☐ cm

☐ cm

☐ cm

☐ cm

☐ cm

☐ cm

☐ cm

☐ cm

(PUPIL PAGE)

Name: _____

The baby frogs' jumping game

Tony jumped $7\frac{1}{2}$ cm

Henry jumped $6\frac{1}{2}$ cm

Gordon jumped 6 cm

Lucy jumped 5 cm

David jumped 3 cm

Estelle jumped 10 cm

Sam jumped $8\frac{1}{2}$ cm

Julie jumped $11\frac{1}{2}$ cm

Classworks © Classworks Numeracy author team, Nelson Thornes Ltd, 2003

Measuring in metres and centimetres

Oral/mental starter p 186

Advance Organiser

We are going to measure length, using metres and centimetres

You will need: metre sticks, tape measures (1 metre long, marked in cm)

Whole-class work

- Measure the width of a table or desk using a metre stick. Try to find a table that has one side longer than 1 metre.

- *This length is longer than a metre, but it is shorter than 2 metres. How can we measure this length more accurately?*

- Demonstrate using a ruler (or metre stick marked with cm) to measure beyond the metre stick. Write the length on the board in the following form: *1m 20 cm*.

- Introduce the term *m* as a short form of metre. *This means 1 metre and 20 centimetres.*

- Measure a few more objects that are longer than 1m (cupboards, shelves, and so on). Try to find an object about one-and-a-half metres long.

- *This length is about 1m and 50 cm long. How else could we write this length?*

- If no one suggests it, point out that there are 100 cm in 1m, and that 50 cm is half a metre. Fold a 1m tape-measure in half to show that half a metre is 50 cm.

- Write various ways of expressing this length on the board, such as 1m 50 cm, 1 metre and 50 centimetres, $1\frac{1}{2}$ m, 1 metre and half a metre.

Independent, paired or group work

- Ask the children, in pairs, to measure three or four appropriate lengths, recording in metres and centimetres each time.

Plenary

- Look at some of the results. *Who can stretch their arms out widest?*

- Use the recorded results to select two or three possible 'widest stretchers'.

- *How can we make sure we are measuring carefully?*

- *What does this length measure? How did you read that off the metre stick?*

Introducing decimal notation

Oral/mental starter
p 186

Advance Organiser

We are going to use decimal notation

You will need: centimetre cubes, metre stick, rulers, resource page E

Whole-class work

- Lay a metre stick on a table wider than 1 metre and place a row of cm cubes along its length. This is not as laborious as you might think – use a couple of rulers to comb the cubes into single rows and into position.

- Ask a couple of children to count the cubes in tens and place a small card marker between each 10.

- Count the cubes in tens to prove that there are 100 cm in a metre.

- Point out that they need to measure longer than 1 metre.

- Write the measurement up on the board in the following form: *1 metre 36 centimetres.*

- Underneath it write: *1m 36 cm.*

- *Can anyone think of an even shorter way to write the same length?*

- Write underneath: *1.36 m.*

- *All of these measurements mean the same thing. They are all equivalent.*

- Practise with some more lengths, writing in one form and asking the children to rewrite in an equivalent form.

Independent, paired or group work

- Ask the children to complete resource page E.

Plenary

- Write the last example in each section of the resource page on the board and discuss. *Who has an answer for these?*

- *Who can tell me an equivalent way of writing 0 m 75 cm?*

- *How do you know? Has anyone any different ways?*

Name: _____

Writing decimal lengths

Write down an equivalent way of expressing each length as in the example.

2 m 35 cm can be written as _____2.35 m_____

5 m 29 cm can be written as _____·_____

28 m 75 cm can be written as _____·_____

45 m 0 cm can be written as _____·_____

125 m 50 cm can be written as _____·_____

0 m 75 cm can be written as _____·_____

Write these lengths as metres and centimetres.

10.86 m can be written as _____10 m 86 cm_____

105.25 m can be written as _____

21.50 m can be written as _____

9.99 m can be written as _____

0.95 m can be written as _____

8.05 m can be written as _____

Classworks © Classworks Numeracy author team, Nelson Thornes Ltd, 2003

Measures (2)

Outcome

Children will be able to use units of time appropriately

Medium-term plan objectives	• Read and begin to write the vocabulary related to time.
	• Use units of time and the relationship between them.

Overview	• Understand relationships between units of time and using units appropriately.

How you could plan this unit

	Stage 1	Stage 2	Stage 3	Stage 4	Stage 5
Content and vocabulary	Units of time *day, week, month, year, hour, minute, second*	More units of time *fortnight, century*			
Notes	Resource pages A and B	Introduce century as a larger unit of time, 100 years. Talk about what you might measure in very large units of time. Introduce the fortnight as two weeks, or fourteen nights.			

Units of time

Advance Organiser

We are going to look at which units of time we can use

Oral/mental starter p 186

You will need: clocks, calendars and other time-measurement devices, resource pages A and B

Whole-class work

- Show the children examples of ways we measure time; for example, the objects mentioned above and the examples on resource page A.
- *Who can tell me what this is?*
- *What units of time will this measure?*
- Write the names of units of time on the board as they are suggested: *second, day, week, year, month, minute, hour.*
- Discuss the idea of measuring time.
- *What other things can we measure?*
- Draw out that we can measure things like length, capacity, weight, as well as time.
- *What units do we use to measure length?*
- *What units do we use to measure capacity?*
- *What units do we use to measure time?*
- Discuss equipment and the units used to measure all of these things.
- *Which units of time might you use to measure how long it is until Christmas?*
- *Why wouldn't you use seconds/minutes, or hours/years?*
- *What equipment would you use to work out the time until Christmas?*

Independent, paired or group work

- Ask the children to complete resource page B.
- Ask them then to write down something they would most sensibly measure in years, weeks, days, hours, minutes and seconds.

Plenary

- Look at some of the children's answers on resource page B.
- *Did everyone have the same answer?*
- *Can anyone think of a different answer?*
- Discuss the children's suggestions for what they would measure with each unit.
- *What else could you measure in years?*
- *What else could you measure in days?*
- *What else could you measure in minutes?*

Name: _____

Time

Name: _____

Units of time

Circle the correct unit each time.

One year is the same as:

12	seconds	minutes	hours	days	weeks	months
52	seconds	minutes	hours	days	weeks	months
365	seconds	minutes	hours	days	weeks	months

One week is the same as _____ days.

One day is the same as _____ hours.

One hour is the same as _____ minutes.

One minute is the same as _____ seconds.

Write these units of time in order starting with the unit of shortest duration:

second   minute   hour   day   week   month

How many minutes are there in half an hour?   _____

How many hours are there in half a day?   _____

How many seconds are there in half a minute?   _____

How many seconds are there in two minutes?   _____

How many weeks are there in half a year?   _____

Classworks © Classworks Numeracy author team, Nelson Thornes Ltd, 2003

Measures (3)

Outcome

Children will be able to read and interpret measuring scales

Medium-term plan objectives

- Read time to 5 minutes on analogue and 12-hour digital clocks (for example, 9:40).

- Read and begin to write the vocabulary related to mass.

- Measure and compare using kilograms and grams, and know the relationship between them.

- Suggest suitable units and equipment to estimate or measure mass.

- Read scales.

- Record measurements using mixed units, or to the nearest whole/half unit (for example, 3.5kg).

- Choose appropriate number operations and calculation methods to solve measurement word problems with one or more steps.

- Explain and record method.

Overview

- Read time to 5 minutes.

- Measure and compare mass using kilograms and grams.

- Use mixed units and round to nearest half kilogram.

How you could plan this unit

	Stage 1	Stage 2	Stage 3	Stage 4	Stage 5
Content and vocabulary	Time to five minutes *o'clock, half past, quarter to, quarter past, minute, hour, second, clock, watch, hands, digital, analogue*	Kilograms and grams *weigh, weighs, balances, kilogram, gram, scales, weight*	More than a kilogram *half kilogram, heavy, light, heavier, lighter*	Reading scales *measure, measuring scale, division, nearly, roughly, about, close to, about the same as, approximately*	
Notes	Resource pages A and B			Resource page C	

Time to five minutes

Advance Organiser

We will tell the time, counting the minutes in fives

**Oral/mental starter
p 186**

You will need: analogue clock face with geared hands and minutes marked in fives, digital clock face (if available), resource page A, resource page B (one per child and one enlarged)

Whole-class work

- Remind the children there are 60 minutes in an hour.

- Show the children an analogue clock face.

- Remind them that each number stands for 5 minutes.

- Point to hour numerals 12 to 6 at random and have the children call out how many minutes past the hour.

- *Does anyone know another way of saying '55 minutes past 2?'*

- *How long is it until 3 o'clock? How do you know?*

- *How could we work it out?*

- Demonstrate counting on in fives from 55, 50, 45, 40 and 35 minutes.

- Point to hour numerals 7 to 11 at random and have the children call out how many minutes to the hour.

- Remind the children that 15, 30 and 45 minutes past are usually known as quarter past, half past and quarter to.

- Set the clock face to a few 'fives' times and record the time on a board in words and in numbers.

- Include: *twenty-five minutes past seven, 7.25,* and so on.

- Include: *o'clock, to and past, half past, quarter to* and *quarter past*.

- Point out that when recording time in numbers we always count the minutes past; but when using words we often use 'to' if it is close to the next hour.

- In turn, ask one child to set the analogue clock face to a 'fives' time and another to set the digital clock face to the same time.

- Repeat, this time with the first child setting the digital and the second the analogue.

Independent, paired or group work

- Ask the children to complete resource pages A and B.

Plenary

- Discuss some of the work from the resource pages.

- Go through a display version of resource page B with the children, inviting them to insert the missing times.

Name: _____

Minutes past

Draw lines to match the pairs of clocks that say the same time.

Classworks © Classworks Numeracy author team, Nelson Thornes Ltd, 2003

Name: _____

Making time

Draw hands to make the analogue clocks match the digital clocks.

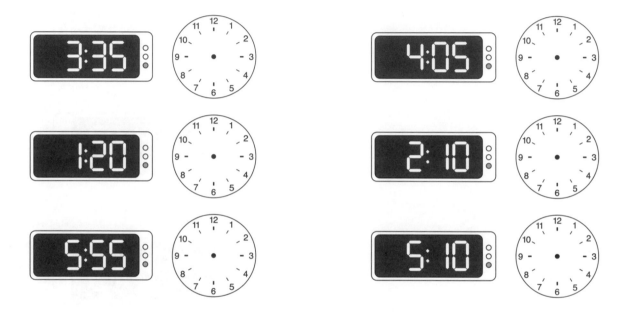

Write times to make the digital clocks match the analogue clocks.

Classworks © Classworks Numeracy author team, Nelson Thornes Ltd, 2003

Kilograms and grams

Advance Organiser

We are going to weigh things using grams and kilograms

Oral/mental starter p 186

You will need: electronic digital scales, Centicubes (many joined together in tens and hundreds), dry sand, packet of crisps, pan balances, kilo masses

Whole-class work

- Show the children a 1kg weight.
- *What could you measure the weight of using kilograms?*
- *What would be difficult to measure in kilograms?*
- Show the class the packet of crisps.
- Demonstrate placing it on the pan balance with the kilogram.
- Show the weight marked on the pack (usually between 25 and 30 grams).
- *Can anyone tell me what a gram is?*
- *Can anyone think of something else you could measure in grams?*
- Write on the board: *kilograms and grams and the abbreviations kg and g.*
- Write: *1000 grams is the same as 1 kilogram.*
- Weigh one Centicube (or two if your balance counts in twos and then split them).
- Give some children the chance to hold the cube and feel how light it is.
- *Can anyone think of things that weigh 20 grams? How about 100 grams?*
- Make ten Centicubes into a rod and join it to nine other 10-rods to make a 10 × 10 flat.
- Point out, if necessary, that the flat has 100 cubes.
- Weigh it – it should be close to 100 g.
- Put ten flats together to form a 10 × 10 × 10 cube and weigh it – it should be close to 1kg.
- Allow some children to hold each weight in their hands.
- They can compare the cube to the kilo weight.
- If there is a discrepancy, ask the children why this might be.
- For example, the scales could be inaccurate, the cubes may not be exact, and so on.

Independent, paired or group work

- Ask the children to find objects to match each of the following weights: 50 g, 100 g, 200 g, 500 g.
- They should record their estimates with an adult, then, they should go in groups to check their estimates in turn and record the actual weights.

Plenary

- Look at some of the decisions made for estimating.
- *Who managed to get nearly right? Who got the closest?*
- *Can you tell us how you made your estimate?*

More than a kilogram

Advance Organiser

We are going to weigh things using both kilograms and grams

Oral/mental starter p 186

You will need: assortment of items to weigh that are heavier than 1 kg (house brick, half brick, made up and labelled 'packets' [containing sand], radio, cassette player, and so on), pan balance

Whole-class work

- Place a heavy object (for example, weighing about two and a half kilos) on a balance and try measuring its weight using just kilogram weights.

- *What is the weight of the radio?*

- Draw out that it is heavier than 2 kg, but lighter than 3 kg.

- *It is between 2 kg and 3 kg.*

- *Can we tell if it is closer to 2 or 3, or if it is exactly half way?*

- *How could we check?*

- Ask how many grams are the same as (equivalent to) half a kilogram.

- Draw out or explain that 500 g is equivalent to half a kilogram.

- Show them a 500 g weight and place it on the balance pan.

- *Do the pans balance now?*

- *How can we make them balance?*

- *Do we need to take or add smaller masses?*

- *Do we need to replace the 500 g mass with a smaller one?*

- Involve the class in using trial-and-improvement in this way, introducing smaller masses to get an approximate measure of mass.

- On the board, record this in the form: *2 kg and 400 g.*

- *Who can suggest another way of writing this?*

- Write on the sheet: *2 kg 400 g.*

Independent, paired or group work

- Ask the children, in groups, to find and weigh four items.

- Remind them that 500 g can be recorded as $\frac{1}{2}$ kg.

- Also remind them, if necessary, that they need to write in the units of measurement.

Plenary

- Look at a few of the objects measured by the children.

- Where the object has been measured by more than one child, compare results.

- Find one with a discrepancy and demonstrate with the class, allowing them to suggest adding in or taking out weights.

Reading scales

Advance Organiser

We are going to look at how to read different sorts of scale

Oral/mental starter p 186

You will need: assorted devices with scales (kitchen and bathroom scales, a clock face, a thermometer, a radio [dial], a tape measure, and so on), resource page C (one per child and one enlarged)

Whole-class work

- Look at a variety of measuring scales.

- *Does anyone know what a scale is?*

- Discuss balance scales, that we are talking about a measuring scale, and that everything you can read a measure from has a measuring scale.

- *What do we read on this scale?* (Weight, temperature, time, radio station/ frequency, length, and so on.)

- *Who can show me how to read the weight from this scale?*

- *Who can demonstrate reading the time from this scale?*

- *How did you do that?*

- Ask the children to talk about the unlabelled markings on some of the scales.

- *What do you think this part of the scale could mean?*

- Demonstrate, if necessary, counting up the divisions between numbered markings on a scale.

- *This scale has ten divisions between 10 and 20.*

- *Count with me: 11, 12, 13, 14, 15, 16, 17, 18, 19, 20.*

- Repeat for other divisions: 100, 110, 120, and so on up to 200.

Independent, paired or group work

- Ask the children to look carefully at the scales on resource sheet C.

- They must be careful to look at the unit each one is measuring in and write the unit on the sheet as well.

- On the temperature gauge, ask children to draw a red line at 25°C, a green line at 17°C, a black line at 46°C and a blue line at –6°C.

Plenary

- Look at an enlarged version of resource page C.

- Check through the readings on the first three scales.

- Now, with the help of the class, draw on the four pointers for the temperature gauge.

- *Starting on the left, which order do the coloured lines come in? (blue, green, red, black)*

- *Was anyone confused by the minus sign on the gauge?*

- Point to –3.

- *What does this reading mean?*

- Explain that this part of the scale means 'minus three degrees Celsius'.

Name: _____

Scales

Read and write down what these scales are showing.

This scale reads _____.

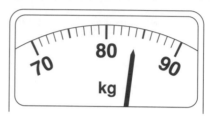

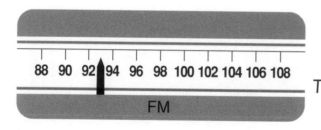

The radio is tuned to _____ FM.

These kitchen scales are showing _____.

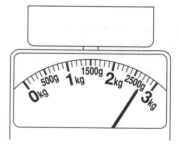

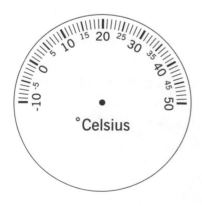

On the temperature gauge draw:
a red needle to show 25°C
a green needle to show 17°C
a black needle to show 46°C
a blue needle to show −6°C.

Measures (4)

Outcome

Children will be able to compare and measure capacity, and solve problems

Medium-term plan objectives	• Read and begin to write the vocabulary related to capacity.
	• Measure and compare using litres and millilitres, and know the relationship between them.
	• Suggest suitable units and equipment to estimate or measure capacity.
	• Read scales.
	• Record measurements using mixed units, or to the nearest whole/half unit (for example, 3.5 litres).
	• Choose appropriate number operations and calculation methods to solve measurement word problems with one or more steps.
	• Explain and record method.
Overview	• Measure the capacity of various containers and use this to compare.
	• Look at suitable measurement methods and select appropriate units.
	• Solve problems related to capacity.

How you could plan this unit

	Stage 1	Stage 2	Stage 3	Stage 4	Stage 5
Content and vocabulary	Introducing millilitres (ml) *capacity, holds, contains, litre, millilitre, container*	Introducing half litres *half litre, full, half full, empty*	Choosing measuring equipment	Capacity problems *measure, calculate, calculation, mental calculation, method, jotting, answer, right, correct, wrong, what could we try next?, how did you work it out?*	
Notes		Make half litre measure (from a half litre milk carton) and a litre measure. Demonstrate filling the half litre from the litre measure to link the half-full litre to the full half litre.	Resource page A	Resource page B	

Introducing millilitres (ml)

Advance Organiser

We are going to measure capacity in litres and millilitres

Oral/mental starter p 186

You will need: soft drinks containers (some labelled in millilitres, some without labels), 1 litre Tetrapaks, plasticine, millilitre measure (made by pressing centimetre cube into plasticine), measuring jugs/jars, water

Whole-class work

- Show the children a 1-litre Tetrapak (as used for fruit juice, milk and so on).
- *What capacity do you think this pack holds? Why do you think that?*
- Write on the board: *1 litre.*
- Compare a 1-millilitre measure with the Tetrapak.
- *How many times would I have to use this small measure to fill the litre measure?*
- After a few guesses, tell the children that the small measure is called 1 millilitre.
- *Does anyone know what 'millilitre' means?*
- Write on the board: *one thousand millilitres are the same as one litre and 1000 ml = 1l.*
- *I would have to use the millilitre measure 1000 times to fill the litre measure!*
- Look at the labels of some soft drink containers.
- *This can has 330 ml in it. It would fill this tiny plasticine measure 330 times!*
- Pour water from a can into a measuring jug/jar and demonstrate reading off the contents in millilitres, or litres and millilitres if appropriate.
- Ask the children to the front of the class to check your reading.

Independent, paired or group work

- Ask the children to write down the capacities of four labelled containers.
- They should then write if this is less than a litre or more than a litre.
- Encourage the children, if appropriate, to write *a lot less than*, *just less than, about the same as,* and so on.
- Ask the children to select a few unlabelled measuring containers and write estimates of their capacity.
- When they have completed their estimates, they should measure their capacity with adult help, carefully reading from the measuring jar each time.

Plenary

- *How well did you do with your estimates? Did anyone get an estimate exactly right? Can anyone show me a very good estimate? How did you make your estimate? Can anyone show me a result that was surprising?*
- Ask some children to the front of the class to explain and demonstrate how to read capacity carefully.

Choosing measuring equipment

Advance Organiser

We are going to look at different ways of measuring capacity

**Oral/mental starter
p 186**

You will need: assortment of measuring containers of varying size (spoons, jugs, buckets, watering cans and so on), resource page A (one per child)

Whole-class work

- Look at a range of measuring containers.

- Talk about things that need their capacity to be measured in small quantities.

- *What might we want to be really extra careful about measuring?*

- Some examples might include medicine, chemicals, and so on.

- Write them on the board.

- *What could we use to measure medicine?*

- *What units could we measure in?*

- *What might we need to measure in very large amounts?*

- Talk about things that could be measured in larger quantities (water in a bath, fuel in a car, large bottles of pop, water to wash a car, and so on).

- *What could you use to measure fuel for a space ship?*

- *What units could you measure the fuel in?*

- Show the children some measuring containers, from small teaspoons to buckets.

- *What could I measure with this? Why do you think that?*

- Talk about the measuring devices you could use to measure small and large quantities.

Independent, paired or group work

- Brainstorm uses for the measuring containers such as a syringe, a medicine spoon, a 750 ml jug and so on.

- Ask the children to complete resource page A.

- They should try to think of the most sensible and realistic answer for each question.

Plenary

- Look at some of the suggestions for what to measure, and in which unit of measurement.

- *How is petrol measured? Does anyone have anything different?*

- Ask what substances need to be measured accurately – ingredients for recipes, medicines and so on.

- Ask what substances do not need to be measured exactly; for example, how much water to put in a bath.

Name: _____

Litres or millilitres

Ring which **unit** (litre or millilitre) you would use to measure:

1	the sugar to put in a cup of tea.	litre	millilitre
2	how much squash to put in a glass.	litre	millilitre
3	the petrol in a car.	litre	millilitre
4	a drink for a pet hamster.	litre	millilitre
5	a drink for a pet elephant.	litre	millilitre
6	the water in a fish tank.	litre	millilitre
7	the ink in a pen.	litre	millilitre
8	the water in a swimming pool.	litre	millilitre
9	the pop in a drink can.	litre	millilitre
10	the water to make a jelly.	litre	millilitre
11	the pop in a large bottle.	litre	millilitre
12	the perfume in a bottle.	litre	millilitre

Capacity problems

Advance Organiser

We are going to solve problems involving capacity

Oral/mental starter p 186

You will need: resource page B (one per child and one enlarged)

Whole-class work

- Write the following problem on the board: *The school cook needs 35 litres of milk to make custard. The milkman delivers 40 litres. How many litres are left after making the custard?* Read the problem together.

- Ask the children to close their eyes and create a picture of the problem in their heads.

- Read the problem slowly. Help them by suggesting the actions that take place at each stage.

- *There are 40 litres of milk and the cook takes 35 litres. We want to know how many are left.*

- This assists children to see that the problem involves subtraction.

- Discuss how the answer can be worked out by breaking it down into stages.

- Write the stages on the board.

- *How many litres are needed (35)?*

- *How much milk does she have (40)?*

- *So, we need to find out how much is left over after you take 35 from 40.*

- *What calculation do we need to do? (40 litres – 35 litres)*

- *Who can help me do this subtraction?*

- *How can we check our answer?*

- Rework the problem with different data (for example, less custard, more milk delivered).

> What we know:
> What we need to find out:
> How we could work it out:
> What calculations we need:
> Working:
> Our answer:

Independent, paired or group work

- Read through each problem on resource page B. Remind the children to close their eyes and picture the scene and what happens.

- Point out the questions from the board that will help them to think about what to do.

- Ask the children to complete resource page B. Point out that they can use the spaces on the sheet to work out their answers.

- When they have completed the activity sheet they may make up problems of their own.

Plenary

- Take each of the problems in turn. Ask for answers.

- If a correct answer is offered, ask the child to explain how they got the answer and to show you their working.

- Work out the answer together using the working-out space of the enlarged version of the activity page.

- Invite the children to suggest problems of their own, or variations of the problems on the sheet.

Name: _____

Capacity

A racing car needs 140 litres of fuel to reach the end of a race.
It starts the race with 80 litres of fuel in its 100-litre tank.
How much more fuel *could* have fitted in?

Halfway through the race the car stops to refuel.
How much fuel is left in the tank?
How much more fuel do they need to put in?

```
This space is for you to work out your answer
```

An elephant in the zoo needs to drink 160 litres of water a day.
The elephant likes to squirt half the water at visiting school children.
How many 10-litre buckets of water does the zookeeper need to fetch?

```
This space is for you to work out your answer
```

For Sports Day each of the 200 children is given a 250 ml drink.
How many litres are drunk?

The mums mix one litre of squash with every 9 litres of water.
How much squash do they use?
How much water do they use?

```
This space is for you to work out your answer
```

Classworks © Classworks Numeracy author team, Nelson Thornes Ltd, 2003

Measures (5)

Outcome

Children will learn how to solve problems using dates

Medium-term plan objectives

- Use a calendar.
- Choose appropriate number operations and calculation methods to solve time word problems with one or two steps.

Overview

- Interpreting a calendar and counting on and back to solve problems.
- Using calculation to work out time intervals across different months.

How you could plan this unit

	Stage 1	Stage 2	Stage 3	Stage 4	Stage 5
Content and vocabulary	Calculating with dates *calendar, date, January, February (and so on), day, week, month, year*	Problems with dates *birthday, holiday*			
Notes		Resource page A			

Calculating with dates

Oral/mental starter p 186

Advance Organiser

We are going to find how many days between two dates

You will need: list of children's dates of birth, calendar pages for May, June, July

Whole-class work

- Write on the board: *date*.

- *Who can tell me what this means?*

- Establish that a date is a specific day, identified by its position in the month and year; for example, 12th August, or 18th February 1975.

- *Who can give me examples of special dates that they know?*

- Write some on the board in the following form: *11th November 2002*.

- *Who can tell me their date of birth?*

- Use your list to check and to add those dates that children are not sure of.

- Read through the dates with the children.

- Look at the part of the calendar for May, June and July.

- Count up how many complete (Sunday to Saturday) weeks there are in the months of May, June and July.

- Add together the days of the part weeks at the start of May and the end of July.

- Pick a day in the first week of a month and count in sevens to establish the dates of that day within the same month.

- Repeat, using a different starting day/date.

Independent, paired or group work

- Ask the children to work out the answers to the following problems.

- *How many Mondays in June? How many Sundays in May? What day is the 11th of July?*

- *Write down the dates of each Saturday in May. Write the date five weeks before 20th June.*

- *How many days are there altogether in May, June and July? What day is 11 weeks after 7th May?*

Plenary

- Look again at the calendar.

- Practise counting in sevens, starting at 7.

- *Does anyone know a mental method for adding seven to any number?*

- Demonstrate adding 7 by partitioning the 7 to make 10 and adding the remainder.

- Discuss what day particular dates will be next year, and why.

Problems with dates

Oral/mental starter
p 186

Advance Organiser

We are going to solve some calendar problems without using a calendar!

You will need: resource page A (enlarged)

Whole-class work

- Ask the children to tell you the months of the year, beginning with January.
- Write them on the board, in order.
- Ask the children to write them in their books at the same time.
- Read slowly through the poem 'Thirty days hath September'.
- Ask the children to write the number of days beside each month in their books as you do so.
- *How many weeks is it from 8th March to 5th April?*
- Take suggestions of how to break down the problem.
- *We could work it out in days first, then work out how many weeks this is.*
- *How many days from the 8th March until the last day of March?*
- *How many days from the last day of March to the 5th April?*
- One way is to use a time line like the one on resource page A. Demonstrate on the board.
- *There are 28 days. How many weeks is that?*
- Encourage or suggest the idea of dividing by the number of days in the week.
- Ask the children to help you divide 28 by 7.
- Try a similar problem; for example, 30th October to 12th December.
- This time work out the days left in October, add on all 30 days of November, then add on the first 12 days of December.
- *It is 43 days. Who can estimate how many weeks that is? Six weeks is 42 days and 1 more day is 43 days. So 43 days is the same as 6 weeks and 1 day.*

Independent, paired or group work

- Ask the children to calculate how many days there are between 20th April and 25th May; 2nd September and 28th October; 12th October and 30th November; 23rd July and 13th August; 24th March and 12th May; 8th June and 24th August; 27th April and 13th July; 4th September and 25th December.

Plenary

- Check through the answers to some of the questions.
- *Did anyone use any shortcuts to work out part of the answer in their head?*
- Make up and solve a problem, with the class, that spans February.

Name: _____

Calculating time intervals

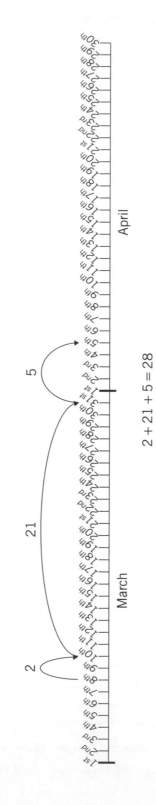

$2 + 21 + 5 = 28$

Shape and Space (1)

Outcome

Children will be able to sort and investigate shapes, identify right angles and locate objects on a grid

Medium-term plan objectives	• Classify and describe 3-D and 2-D shapes, referring to reflective symmetry, faces, sides/edges, vertices, angles.
	• Read and begin to write the vocabulary of position.
	• Use spaces on square grids.
	• Identify right-angles in 2-D shapes and in the environment.
	• Investigate general statements about shapes.
Overview	• Sort quadrilaterals using a Venn diagram.
	• Investigate right-angles in the environment.
	• Locate spaces in square grids.
	• Investigate general statements about shapes.

How you could plan this unit

	Stage 1	Stage 2	Stage 3	Stage 4	Stage 5
Content and vocabulary	Sorting quadrilaterals square, rectangle, rectangular, quadrilateral, flat, side, sort, right-angle	Turning right angles clockwise, anti-clockwise, movement, whole turn, half turn, quarter turn, right-angle	Identifying spaces in square grids map, plan, left, right, up, down, across, along, grid, row, column, horizontal, vertical	Investigating statements about 3-D shapes pyramid, face, edge, sort, pattern, method, what could we try next?, how did you work it out?	
Notes		Resource page A	Resource page B		

Sorting quadrilaterals

Oral/mental starter
pp 187–188

Advance Organiser

We are going to use a Venn diagram to sort quadrilaterals into sets

You will need: Blu-Tack, selection of four-sided 2-D shapes

Whole-class work

● Write the following word on the board: *quadrilateral*.

● *Can anyone tell me what this means?*

● Some children may have heard other words with quad in, such as quadrant or quad-bike.

● *'Quad' means four. A quadrilateral is a shape with four sides.*

● Draw a Venn diagram on the board:

● *Who can find a shape that belongs in this set?*

● Revise right-angles, if necessary.

● Ask a child to find a shape and use Blu-Tack to stick it to the board.

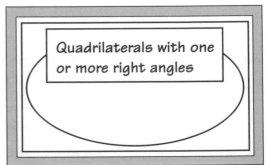

Quadrilaterals with one or more right angles

● *Who else thinks that this shape belongs in this set?*

● *Can you tell me why you think that?*

● Look at the remaining shapes. *Do any have right-angles?*

● Test the angle against an agreed right-angle in the classroom, such as the corner of a sheet of paper or the corner of a table.

● Continue to ask the children to use Blu-Tack to stick quadrilaterals inside the set.

● Keep getting the class to agree or disagree with each decision, and explain why.

● *Look at our set. Who can tell me why the oval enclosure is inside the large rectangular enclosure, not outside it?*

Independent, paired or group work

● Give the children their own set of shapes.

● Change the label of the subset to Quadrilaterals with four right angles. Ask them to use the diagram to sort their shapes.

Plenary

● *Which shapes were in the 'quadrilaterals' set?*

● *Which shapes had the property 'all angles are right-angles'?*

● *Show me on your Venn diagram.*

● *What if the property was 'quadrilaterals with no right-angles'?*

● *Which shapes would be in that set?*

● Draw a Venn diagram to represent this set and subset.

● Ask the class to help you place the shapes in the correct section.

Turning right angles

Advance Organiser

We are going to show right-angle turns using pointers

Oral/mental starter pp 187–188

You will need: resource page A (one per pupil and one enlarged), a dial with two pointers of equal length, paper fastener

Whole-class work

- Ask some children to make different right-angles using their arms.
- The other children say whether it is a right-angle or not, and why.
- If necessary, use a book to check the angle.
- Show the children a dial with both pointers at 0 (made with a paper fastener).
- *I am going to move one of the pointers to make a right-angle.*
- *To which number should it point?*
- Take suggestions from the children and discuss them.
- Ask a child to move one of the pointers to make a right-angle as discussed.
- *Were we right? Who can tell me another number that would make a right-angle?*
- Repeat, starting with both pointers on 1.

- *Which number should I move the pointer to now? How do you know?*
- Repeat with other numbers, asking the children each time to predict and turn the pointers to make a right-angle.
- Show the children an enlarged version of resource page A.
- Discuss the different dials and demonstrate recording a right-angle on the sheet.

Independent, paired or group work

- Give each child a copy of resource page A.
- Ask them to use the numbers to draw as many different right-angles on dials as possible.

Plenary

- Ask one child to explain how many different right-angles they found and how they knew they had not missed any out.
- Involve the children in a discussion of how you can test a right-angle.
- End by asking the children to use their arms to demonstrate a right-angle.

Name: _____

Dials

Identifying spaces in square grids

Advance Organiser

We are going to find objects on a grid

Oral/mental starter pp 187–188

You will need: resource page B (one per child and one enlarged), small models (or pictures) of trees, bridges etc.

Whole-class work

- Use the first map grid from resource page B.

- *Can anyone tell me what this is? Where have you seen one before?*

- If necessary, point out that the columns are labelled with letters and the rows with numbers.

- Ask someone to explain the difference between a column and a row.

- *A column is vertical and a row is horizontal.*

- Point out, if no one else does, that the letters and numbers refer to squares rather than grid points.

- Practise finding particular squares by asking children to locate; for example, E5, G1, and so on.

- Read the instructions for placing features on the map on resource page B.

- Ask different children to draw items in the correct places.

- Encourage the children to find the letter first, then move up the column until they reach the correct row.

- Build up the picture of the map. Think of some more items you could add. Decide upon a location for the item. Ask someone to draw it.

Independent, paired or group work

- Give each child a copy of resource page B.

- Ask the children to complete the map and add two items.

- They should write the name and position of the new items on the list.

- *Invent your own items and locations for the second map.*

- *Make a list of locations without filling in the map.*

- *Exchange with a partner to complete each other's maps.*

Plenary

- Pick out some items and ask the children to say the location of each.

- *What did you add to the map?*

- *Can you tell me which square it is on?*

- Confirm that the references to squares start with a letter (for the column) and then have a number (for the row).

- *First we read across, then up.*

(PUPIL PAGE)

Name: _____

Spaces on a grid

Draw these on the map:

a hut at D5

a dragon at F2

a cave at G3

trees at E2, E3, D2, D3, D4

a river from F7 through E6, D6, C5, B5 to A4

a bridge at C5

a path from A7 through B6, C5 to D5

Write your own instructions.

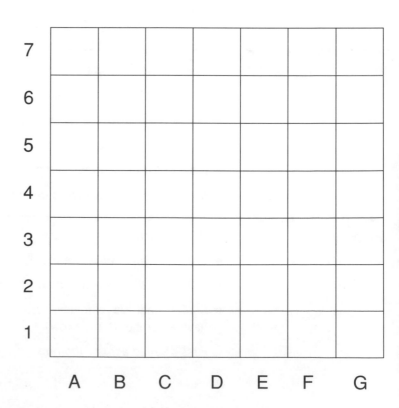

Classworks © Classworks Numeracy author team, Nelson Thornes Ltd, 2003

Investigating statements about 3-D shapes

Advance Organiser

We are going to find out if the number of edges of a pyramid is always even

Oral/mental starter pp 187–188

You will need: as many pyramids as possible (wooden or plastic shapes, Polydron, Clixi, including, if possible, pentagonal-based, hexagonal-based and similar 'unusual' pyramids)

Whole-class work

- Hold up a pyramid. *Who can tell me the name of this shape?*

- *Can anyone tell me its properties? Does anyone know any other sorts of pyramid?*

- Write up suggestions on the board and verify them with the class.

- Once they (or you) have suggested at least tetrahedrons and square-based pyramids, ask: *How many edges are there on a tetrahedron? (six) How many on a square-based pyramid? (eight)*

- *I think that every pyramid has an even number of edges. How can I find out? What other sorts of pyramid can you think of?*

- If necessary, draw the children's attention to the shape at the base of the pyramid.

- Suggest pyramids with different shapes at the base.

- *What if the base were a pentagon? How many faces would the pyramid have?*

- *How many edges?* If possible, show the children these pyramids and count up the edges.

Name of pyramid	Number of edges
tetrahedron	6
square-based	8

- *How can we record our information?* Write the headings for a table on the board.

Independent, paired or group work

- Ask the children to work in pairs or groups to investigate the problem.

- They should record their work in their exercise books using the table on the board as a model.

- Encourage them to think about patterns between the properties – the number of edges compared to the number of faces, the shape of the base compared to the number of edges and so on.

Plenary

- Ask a child to read their own notes.

- *What did you find out? What methods did you use? Did anyone do anything differently? Did anyone notice any patterns?*

- Write the findings on the board.

- Children should have discovered that the number of edges of a pyramid is double the number of faces. Another way of expressing this is that the number of edges of a pyramid is double the number of sides of the shape of the base.

Shape and Space (2)

Outcome

Children will be able to draw solid shapes and investigate patterns in shape and space

Medium-term plan objectives

- Make and describe shapes and patterns.
- Relate solid shapes to pictures of them.
- Read and begin to write the vocabulary of direction.
- Make and use right-angled turns, and use the four compass points.
- Solve shape problems or puzzles.
- Explain reasoning and methods.

Overview

- Make and describe shapes and patterns.
- Relate solid shapes to pictures of them.
- Use the four compass points.
- Solve shape problems or puzzles.

How you could plan this unit

	Stage 1	Stage 2	Stage 3	Stage 4	Stage 5
Content and vocabulary	Stars inside polygons *rectangle, square, star, pentagon, hexagon, octagon, diagonal*	Pictures of solid shapes *cube, cuboid, draw, straight*	Using four compass points *north, south, east, west, N, S, E, W, compass point*	Shapes made with four squares *square, line of symmetry, pattern, mirror line, reflection, rotation*	
Notes		Resource page A	Resource page B	Resource page C	

Stars inside polygons

Advance Organiser

We are going to investigate drawing stars inside polygons

Oral/mental starter pp 187–188

You will need: flat shapes to draw around (including heptagon, octagon, pentagons and hexagons), rulers, pencils

Whole-class work

- Draw a pentagon on the board, as accurately as possible.

- *Who can tell me what a polygon is?*

- *Who can tell me the name of this polygon?*

- *Who can tell me what the word 'diagonal' means?*

- Establish that 'diagonal' means 'a straight line connecting two corners of a polygon'.

- Ask the children to draw some diagonal lines on the pentagon, connecting two corners.

- *How many diagonals can you draw in a rectangle?*

- *How many can you draw in a pentagon?*

- Use a ruler to draw them. (There are five, one from each point to the next point but one.)

- *What shape have we made? What shape is in the centre?*

- Draw the diagonals of the smaller pentagon.

- *What shape do you think we will make in the centre? How do you know?*

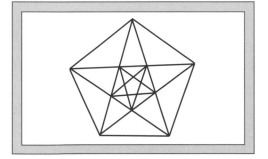

Independent, paired or group work

- Give the children pentagons, hexagons, heptagons and octagons to draw around.

- Ask them to draw diagonals on each shape using a ruler.

- They should then try to continue with any shape formed inside the new shape.

- *Can you make a star inside a hexagon?*

- *Can you make a star inside an octagon?*

Plenary

- *Which polygons make stars? Which do not?*

- *What is special about the polygons that do make stars?*

- Draw out that you can only make stars using this method inside a polygon with an odd number of sides.

Pictures of solid shapes

Oral/mental starter
pp 187–188

Advance Organiser

We are going to use spotted paper to draw pictures of solid shapes

You will need: linking cubes, resource page A (one per child and one enlarged), dotted paper (per child)

Whole-class work

- Show the children an enlarged version of resource page A.
- Show them how to complete the picture of the cube line by line.
- Point out which dots you are joining.
- Ask a child to copy the picture of the cube carefully, in the same way, in front of the class.
- Show the children how to extend the picture of the cube vertically or horizontally to make a cuboid.
- Look at the pattern of 'steps' at the bottom of the page.
- *Can you work out how many cubes would be needed to make this shape?*
- *How did you work that out?*
- Show the children how to copy the shape, line by line.
- Start at the top.
- Draw attention to each line as you draw it and how the dots on the paper are used as a guide.
- Ask a child to make the shape with cubes.

Independent, paired or group work

- Give the children a copy each of resource page A. Ask the children to begin by copying the cube, then extend it to draw a cuboid.
- Next they should copy the steps pattern.
- Then they should use linking cubes to make a 4-cube shape and try to draw it on the spotted paper.
- Finally, they can investigate colouring the faces of their cubes to make patterns.

Plenary

- Look at all the cubes, cuboids and patterns.
- *Which were the most difficult to draw?*
- *Why do you think that was?*
- *How did you know how to draw your shape?*

Name: _____

Drawing solid shapes

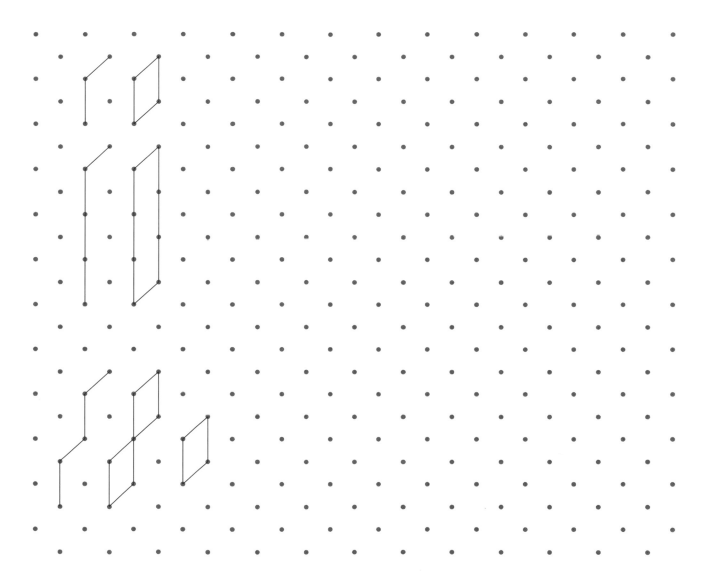

Using four compass points

Oral/mental starter pp 187-188

Advance Organiser

We are going to give directions using north, south, east and west

You will need: 2-cm squared paper cut-out star to fit squares on enlarged resource page B, Blu-Tack, resource page B (enlarged)

Whole-class work

- Show the children an enlarged version of resource page B.

- Ask them to identify north, south, east and west on the sheet.

- Explain that the aim of the game is to write directions for travelling through the maze, collecting at least ten pieces of gold.

- You collect a piece of gold by landing on it exactly at the end of a turn.

- Ask the children in turn to suggest a direction and number of squares; for example, south 4.

- Ask a child to move the star to the correct square each time.

- Write a list of the directions as you make each move on the board.

- Trace the route with a pencil line to show where you are.

- Cross through each piece of gold as it is collected.

- Continue until you have at least ten gold coins, then ask the children to help you get to the finish.

Independent, paired or group work

- Give the children a copy of resource page B and ask them to play the game.

- They should take turns to make and record one move (a move is a direction and a number; for example, east 3).

- If they land on a coin on their move, they should write their initials in that square.

- The winner is the player with most gold coins.

Plenary

- Write 'north' at the top of an enlarged sheet of 2-cm squared paper.

- Place your pen in one of the squares.

- Ask different children to give you a direction to move in and a number of squares to move.

- Draw a pattern as you move.

- *What sort of pattern will I make?*

- *Can you give me directions to make a star?*

Name: _____

Compass directions

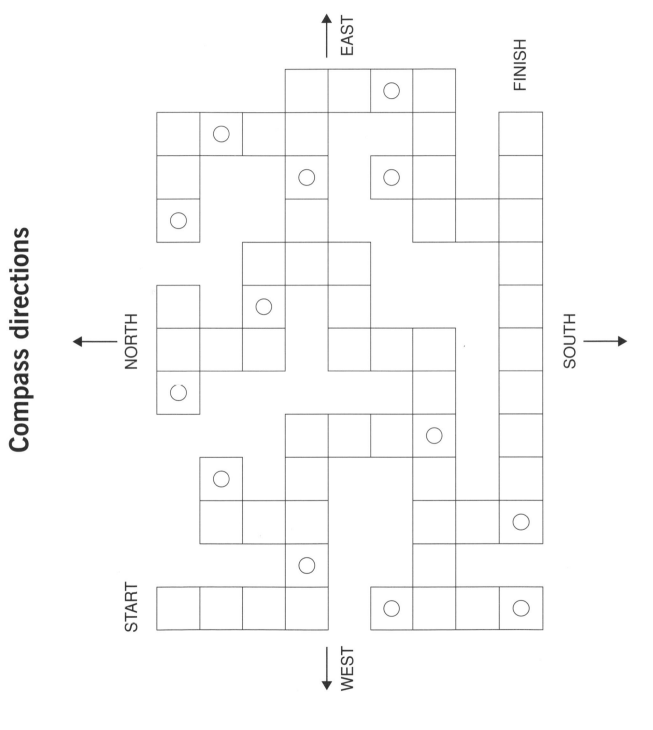

Shapes made with four squares

Advance Organiser

We are going to make as many different shapes as possible using four squares

Oral/mental starter pp 187–188

You will need: linking squares (such as Clixi or Polydron), squared paper, coloured pens or pencils, resource page C (one copy)

Whole-class work

- Ask a child to take four squares and fit them together.

- The rules are that squares can only touch along a side, not just at a corner.

- Ask them to record the shape by drawing the flat shape on the board.

- Ask another child to make a different shape. Ask the child to record it.

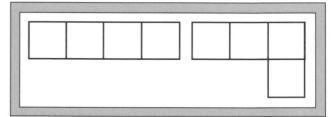

- Continue making different shapes until the children cannot think of any more.

- Look at resource page C for an exhaustive list of shapes made with four squares.

- Look at shapes that are reflections or rotations of each other by asking a different child to draw each shape in a different orientation.

- *Can we see this shape anywhere else? Have we made this shape already?*

- *What would this shape look like the other way up?*

- *What would it look like reflected in a mirror?*

- Erase extra shapes that match in this way. *How could we sort our shapes?*

- Encourage the children to sort the shapes by the number of squares in a line (four, three or two). At an appropriate point, suggest that this could also be a systematic way of finding all the possible shapes.

- *How many shapes can we make with four squares in a row? How many with three squares in a row?*

Independent, paired or group work

- Ask the children to make all the shapes they can from the four squares.

- They should use one colour for each shape.

- They should then try to fit all their shapes together into a rectangle.

Plenary

- Look first at any shapes that have been fitted into a rectangle.

- Check with the class for gaps and that each pattern has four squares.

- Count the number of times each shape was used to check that the class can identify rotations and reflections.

- *Is it easier if you use fewer shapes? Which shapes are easy to fit together? Which shapes make the puzzle difficult? Why is that?*

Name: _____

Shapes that can be made with four squares (rotations not shown)

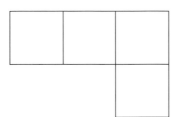

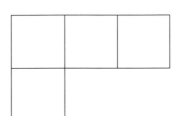

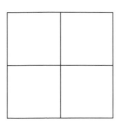

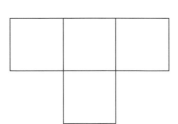

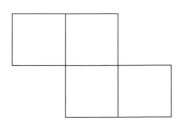

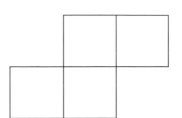

Classworks © Classworks Numeracy author team, Nelson Thornes Ltd, 2003

Shape and Space (3)

Outcome

Children will be able to investigate symmetry, right-angles and tessellation, and use compass directions

Medium-term plan objectives	
	• Identify and sketch lines of symmetry, recognise shapes with no line of symmetry.
	• Sketch reflection of simple shape in a mirror.
	• Read and begin to write the vocabulary of position, direction and movement.
	• Recognise that a straight line is two right-angles.
	• Compare angles with a right-angle, saying whether they are more or less.
	• Investigate general statements about shapes and suggest examples to match them.
	• Explain reasoning.

Overview

- Identify lines of symmetry in flags; identify absence of symmetry.
- Read and write the vocabulary of position, direction and movement.
- Recognise that a straight line is two right-angles.
- Investigate shapes that tessellate.

How you could plan this unit

	Stage 1	Stage 2	Stage 3	Stage 4	Stage 5
Content and vocabulary	Shapes with, and without, symmetry *line of symmetry, symmetrical, fold, match, mirror line, reflection*	Using position, direction and movement *north, south, east, west, N, S, E, W, compass point*	A straight line is two right-angles *straight line, right-angle, rectangle, square*	Investigating shapes that tessellate *pattern, puzzle, what could we try next?, how did you work it out?*	
Notes	Resource page A	Resource page B			

152

Shapes with, and without, symmetry

Advance Organiser

We are going to decide which flags have lines of symmetry

Oral/mental starter pp 187–188

You will need: resource page A (one per child and one enlarged)

Whole-class work

- Look at the flags on the enlarged copy of resource page A.
- *Which ones have no line of symmetry? How can you tell?*
- Demonstrate using a mirror to check.
- Ask a child to choose a flag that is symmetrical.
- *How do you know it is symmetrical?*
- *Where would you draw the line of symmetry?*
- *Is there a second line of symmetry?*
- Ask a child to use the mirror to check.
- Ask a child to colour a flag in the top line so that it has two lines of symmetry.

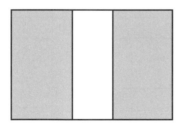

- Ask another child to colour a similar pattern so that it has only one line of symmetry.
- Repeat this with a different pattern to establish how colour can affect the symmetry of the flags.

Independent, paired or group work

- Ask the children to draw lines of symmetry, if any, on the remaining flags.
- They should colour the flags so that each flag retains its symmetry.
- On the three empty flags on the sheet, ask them to design one flag with no line of symmetry, one with one line and one with two lines.

Plenary

- Look at the children's flag designs.
- *Who can help me check the lines of symmetry?*
- Check that they meet the criteria of none, one or two lines of symmetry.

PUPIL PAGE

Name: _____

Lines of symmetry

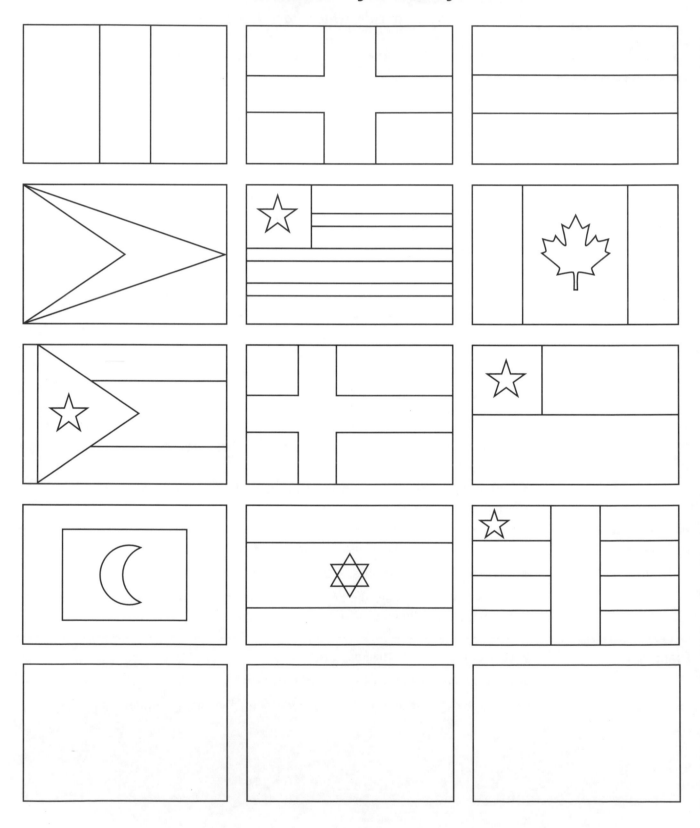

Classworks © Classworks Numeracy author team, Nelson Thornes Ltd, 2003

Using position, direction and movement

Advance Organiser

We are going to read and write directions for travelling around a map

**Oral/mental starter
pp 187–188**

You will need: resource page B (one per child and one enlarged)

Whole-class work

- Locate and read the four compass points on the map.
- *Can anyone tell me what these are?*
- *What do you know about the points of the compass?*
- Ask the children to identify various features on the map.
- *What is in the top right-hand corner?*
- Show the children a magnetic compass to demonstrate the idea of north remaining constant, irrespective of the way you are facing.
- Read the first two lines of the story on the sheet.
- Ask a child to draw the route moving in straight lines, then write the compass direction.
- They can only move north, south, east or west.
- That child chooses another child to read and record the next part of the journey.
- When the whole route has been drawn, check each direction by retracing the route from the beginning.

Independent, paired or group work

- Ask the children to write a similar story, using the features on the map.
- They have to travel in straight lines horizontally and vertically on the map.
- This means they should only use north, south, east and west.
- They then give their map to a partner.
- The partner draws the route as they describe it in the story.
- The first child then checks again.

Plenary

- Read some of the journeys from paired work.
- Use different coloured pens to trace the journeys on the map.
- Ask the children to look out for anything they disagree with and to say why.

Name: _____

Where to go in Bridgetown

After school Anish met his brother outside the Mosque.

Then they went to the High Street.

After that they visited the library.

Then they walked to the lake.

Classworks © Classworks Numeracy author team, Nelson Thornes Ltd, 2003

A straight line is two right-angles

Oral/mental starter
pp 187–188

Advance Organiser

We are going to show that two right-angles make a straight line

You will need: paper for folding

Whole-class work

- Begin two lines of squares on the board as shown below.
- Ask a child to mark all the right-angles. Continue adding squares.
- Ask a child to mark all the right-angles each time.

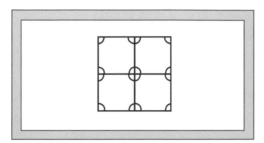

- Find all the places where two right-angles make a straight line.
- Colour each pair of angles a different colour.
- *What will happen if we begin a second line?*
- Point out that two right-angles make horizontal and vertical lines in the pattern.
- *Does anyone know another way to mark right angles?*
- Introduce a special way to mark right angles and replace the curved angle signs with square right-angle signs. *Why do you think we use this sign?*
- Encourage suggestions that squares and rectangles have this kind of angle.
- Show the children how to mark all the right-angles on both sides of a piece of paper.
- Fold the paper in half and mark all the right-angles on both sides again.
- Continue folding in half and marking right-angles at all the corners.
- Unfold the paper and ask the children to find all the pairs of right-angles that make straight lines.

Independent, paired or group work

- Give each child a piece of paper. Ask them to mark all the right-angles in the same way.
- Ask them to unfold the paper and colour pairs of right-angles that make a straight line.
- *How many pairs did you find?*

Plenary

- Open a book and ask a child to find the places where two right-angles make a straight line.
- Find other examples in the classroom; for example, corners of window panes, corners of adjoining tables, floor tiles and so on.

Investigating shapes that tessellate

Advance Organiser

We are going to find out which shapes fit together without any gaps

Oral/mental starter
pp 187–188

You will need: shape templates, spotted paper, isometric paper and similar

Whole-class work

- Make a list of familiar polygons and record their names and the numbers of sides they have.

- Write them in order of number of sides.

- Show the children the square.

- *If you had lots of squares this size, could you fit them together without leaving any gaps?*

- *Why do you think that?*

- Ask the children to help you find out.

- Demonstrate recording the tessellation on appropriate spotted paper. Repeat for triangles.

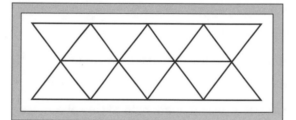

Independent, paired or group work

- Give each group a variety of polygons to fit together.

- They investigate which tessellate, and which do not, and record their work on appropriate spotted paper.

Plenary

- Look at some examples of the children's tiled patterns.

- *Were there any surprises?*

- *Do all shapes with straight sides fit together without gaps?*

- *Do all regular shapes fit together without gaps?*

- *What have we found out?*

Handling Data (1)

Outcome

Children will be able to suggest ways of solving a problem using frequency tables and pictograms

Medium-term plan objectives	• Solve a given problem by organising and interpreting data in frequency tables, and in pictograms with the symbol representing two units.

Overview	• Pose problems.
	• Collect data in a frequency table.
	• Organise into pictograms.
	• Discuss outcome of investigation.

How you could plan this unit

	Stage 1	Stage 2	Stage 3	Stage 4	Stage 5
Content and vocabulary	Interpreting pictograms *pictogram, represent, chart, label, title, axis, axes, most popular, least popular*	Pictograms *count, tally, sort, vote, group, set, frequency table*	Building a pictogram		
Notes	Show the children examples of pictograms. Point out the features. Ask them to discuss the differences and similarities between each one.				

Pictograms

Oral/mental starter
p 188

Advance Organiser

We are going to display information about our birth months in a pictogram

You will need: example of a pictogram

Whole-class work

- *I predict that most of you were born in the first six months of the year.*

- *How can we test if I am right? What do we need to know? How can we find that out?*

- Lead them towards the idea of some form of voting, such as a show of hands.

- Draw a frquency table on the board as shown.

- *Who can tell me what this is?*

- *How can I use this frequency table to record our results?*

- Record the children's birth months one at a time using single tally marks.

Month	Tally	Total
January		

- Add up the totals for each month, counting with the class for some and asking children to count for you for others.

- *Can anyone think of another way of recording this information?*

- Look at an example of a pictogram.

- *Who can tell me something about this chart?*

- Point out the title, axes and labels. Point to the key.

Our birth months

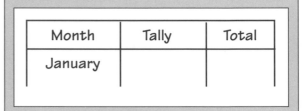

Key: ☺ = 2 children

| January | ☺☾ |
| February | ☺☺ |

- Encourage the children to recognise that the key is telling them how many children are represented by each smiley face.

- Complete two rows of a pictogram for the frequency table on the board, choosing, if possible, a month with an even score and a month with an odd score that will require half a smiley face.

Independent, paired or group work

- Ask the children to copy the frequency table. Encourage them to check by counting the tallies.

- They should then transfer this data to a pictogram using one smiley face to represent two children.

Plenary

- *I predicted that most of you were born in the first six months of the year.*

- *Who can tell me if I was correct? How can you tell?*

- Complete the pictogram on the board, asking children for input and discussion at every stage.

Building a pictogram

Oral/mental starter
p 188

Advance Organiser

We are going to find out our favourite television programme using a pictogram

Whole-class work

- Ask the children to guess at what the class's favourite television programme might be.

- Restrict this to their favourite children's television programme, if necessary.

- Discuss the number of suggestions and suggest narrowing them down to six categories.

- Discuss how to decide these six – you might choose to combine similar programmes (cartoons, football, drama, and so on).

- Draw a tally chart on the board.

- *Where should I write our choices? How can we record which ones we like?*

- Suggest or agree on a vote of some sort and record the children's choices, one at a time, using tally marks.

- Ask the children to help you add the tallies and write the totals in the last column.

- *Who can tell me how we could display our information on a pictogram?*

- Remind the children that you need to write a title, labels on the columns and a key.

- *How can we fit all our data on the chart easily?*

- Agree on using one symbol (for example, a box or circle) to represent two children.

- Record this decision by writing a key on a chart on the board.

- Demonstrate recording one odd and one even score using a half-symbol for the odd score.

Independent, paired or group work

- Ask the children to record the data from the frequency table on a pictogram.

- Remind them to write a title, labels and a key.

- Remind them to use half-symbols to represent one vote.

Plenary

- *Which is our favourite television programme? How can you tell?*

- Complete a pictogram on the board with the class.

- Look at the tally chart and the pictogram.

- *What is different about each chart?*

- *What is easier to see on the tally chart?*

- *What is easier to see on the pictogram?*

Handling Data (2)

Outcome

Children will be able to solve problems by organising and interpreting data in bar charts

Medium-term plan objectives	• Solve a given problem by organising and interpreting data in bar charts – intervals labelled in ones then twos.
Overview	• Pose problems. • Collect data in a frequency table. • Organise into bar charts. • Discuss outcome of investigation.

How you could plan this unit

	Stage 1	Stage 2	Stage 3	Stage 4	Stage 5
Content and vocabulary	Bar charts *count, tally, sort, vote, graph, block graph, bar chart, label, title, axis, axes*	Bar charts (scale marked in twos)			
Notes	Resource page A				

Bar charts

We are going to make bar charts to show how many boys and girls came to school last week

Oral/mental starter p 188

You will need: record of class attendance for each day last week for boys and girls written on the board, resource page A

Whole-class work

- *Is the number of children who come to school different every day?*
- *How can we find out?*
- Discuss various charts and tables that the children are familiar with.
- *Does anyone know a different sort of chart?*
- Show them some examples of bar charts (one is given on resource page A).
- Introduce the term 'bar chart' and write it on the board.
- Draw a set of axes on the board as shown.
- *First I want you to help me fill in this chart for the number of boys who came to school last week.*
- Point out that the title is missing and ask the children to suggest one.
- Demonstrate beginning to complete the bar chart for the boys' attendance.
- Slide your left-hand finger up the scale to the correct value.
- Slide your right-hand finger across the x-axis to find the correct day.
- Move both fingers until they meet at the correct value on the correct column.
- Put a pencil mark there and use a ruler to draw in the sides and top of the bar.
- Colour in the bar.

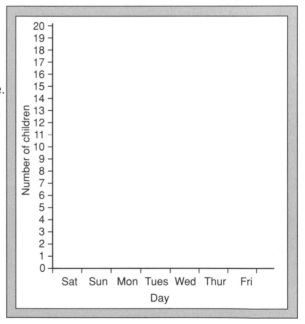

Independent, paired or group work

- Ask the children to copy the axes and make a bar chart for the girls' data.
- They should then write how many girls came to school on Monday, and the days on which most, and fewest, girls came to school.
- Complete the chart for the boys' attendance as they work.

Plenary

- Ask the children to tell you the answers to the questions.
- Discuss the chart on the board as well, then ask questions about the two sets of data.

EXAMPLE

Favourite ice cream flavours

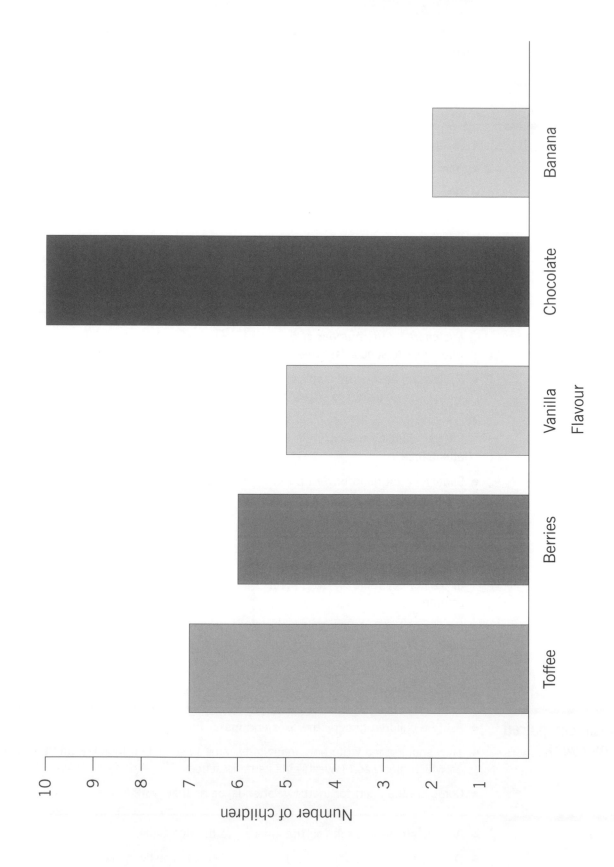

Bar charts (scale marked in twos)

Oral/mental starter
p 188

Advance Organiser

We are going to make a bar chart to show how many children in this class came to school last week

You will need: record of class attendance for each day last week on display

Whole-class work

- *We are going to find out if the numbers of children coming to school last week were different every day.*

- Draw axes on the board as shown.

- *What can you tell me about this chart?*

- Point out the title, axes, labels and vertical scale.

- *What do you notice about the scale?*

- *Which numbers are missing?*

- *How could we mark on odd numbers?*

- Point out that the marks halfway between each number represent the odd numbers.

- Demonstrate filling in the first column on the bar chart based on the data on display.

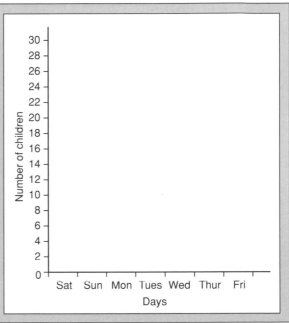

Independent, paired or group work

- Ask the children to use the data on the board to complete their own copy of the bar chart.

- They should answer the questions after completing the chart.

Plenary

- Look at the children's graphs and ask them to help you complete the one on the board.

- Discuss the questions that the children have answered.

- *Does the number of children attending change as the week goes on?*

- *How does it change? Which school day has the fewest?*

- Ask the class to help you list the days in order of attendance (include Saturday and Sunday).

Handling Data (3)

Outcome

Children will be able to solve problems by organising and interpreting data in Venn diagrams and Carroll diagrams

Medium-term plan objectives

- Solve a given problem by organising and interpreting data in Venn and Carroll diagrams – one criterion.

Overview

- Collate data in a Venn diagram.
- Present the same data in a Carroll diagram.
- Compare both diagrams.

How you could plan this unit

	Stage 1	Stage 2	Stage 3	Stage 4	Stage 5
Content and vocabulary	Venn diagrams *count, tally, sort, vote, represent, group, set, list, Venn diagram, diagram, title, label*	Carroll and Venn diagrams *Carroll diagram, interpret, present, represent*			
Notes					

166

Venn diagrams

Oral/mental starter p 188

Advance Organiser

We are going to put data into a Venn diagram

You will need: set of polygons with differing numbers of sides, Blu-Tack

Whole-class work

- Show the children the set of polygons.

- *What is the same about these shapes?*

- *What is different?*

- *How could we group them by their properties?*

- Lead the children to decide on grouping by the number of sides.

- Draw a Venn diagram on the board as shown.

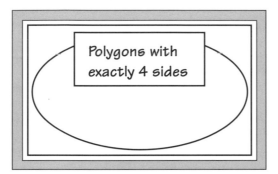

Polygons with exactly 4 sides

- *Who can tell me what this is called?*

- *Can anyone tell me how to use it?*

- Ask the children to use Blu-Tack to stick the shapes onto the correct space on the diagram.

- Conclude that, by default, all the shapes that do not have four sides must be outside the set rectangle but inside the universal set ring.

Independent, paired or group work

- Ask the children, in groups, to complete a Venn diagram for a chosen shape properly; for example, polygons with three sides, with five or more corners, and so on.

Plenary

- Ask the children to tell the class what property they chose. Discuss with the class how to draw that diagram and how to sort the shapes.

- Ask questions about the diagram.

- Discuss the advantages of the Venn diagram as opposed to a simple list.

Carroll and Venn diagrams

Oral/mental starter p 188

Advance Organiser

We are going to use both Carroll and Venn diagrams to solve a problem

You will need: set of polygons with different numbers of sides, Blu-Tack

Whole-class work

- Show the children the set of polygons.

- *What is the same about these shapes? What is different?*

- *How could we group them by their properties?*

- Lead the children to decide on, for example, grouping by number of sides.

- Draw a Carroll diagram on the board as shown.

- *Who can tell me about this diagram? Has anyone seen one before?*

- Confirm that it is called a Carroll diagram (write it on the board).

- Ask some children to come to the board and stick shapes in the correct part of the diagram.

- Now draw a Venn diagram for the same data on the board (see page 167).

- *Who can tell me what this tells us?*
  Discuss the two sorts of diagram

Our class set of polygons	
Has exactly 4 sides	Does not have exactly 4 sides

- Point out that the precise location of shapes is not important, it is the placement within the correct set(s) that matters.

Independent, paired or group work

- Ask the children, in groups, to complete a Carroll diagram for a chosen shape property.

Plenary

- Draw Venn and Carroll diagrams on the board for one group's shape property.

- Discuss the differences between the two ways of representing the data.

- The Venn diagram groups together shapes with the same property.

- The Carroll diagram gives both sets of data equal status side by side.

- *What is clearer about the Carroll diagram? What is clearer about the Venn diagram?*

Fractions (1)

Outcome

Children will be able to draw and write unit fractions of shapes and numbers and recognise non-unit fractions of shapes

| **Medium-term plan objectives** | • Recognise unit fractions $\frac{1}{2}$, $\frac{1}{3}$, $\frac{1}{4}$, $\frac{1}{5}$, $\frac{1}{10}$ and use them to find fractions of shapes and numbers. |
| | • Begin to recognise fractions that are several parts of a whole $\frac{2}{3}$, $\frac{3}{4}$, $\frac{3}{10}$. |

Overview	• Write the fraction of shapes.
	• Find unit fractions of quantities.
	• Identify non-unit fractions of shapes and lines.

How you could plan this unit

	Stage 1	Stage 2	Stage 3	Stage 4	Stage 5
Content and vocabulary	Fractions of shapes *part, equal parts, fraction, one whole, one-half, two-halves, one-quarter, one-third, one-tenth, one-fifth*	Fractions of numbers	Non-unit fractions *two-quarters, three-quarters, four-quarters, two-thirds*		
Notes		Resource page A	Resource page B		

Fractions of shapes

Oral/mental starter
p 184

Advance Organiser

We are going to find one-third of a shape

You will need: linking cubes in different colours (for the children), strips of 2 cm
squared paper, Clixi or Polydron

Whole-class work

- Build 'trains' of linking cubes with the children's help. Begin by making a whole train (for example, of four cubes) in one colour as a reference point and then ask the children to build a matching train using a distinctive colour for the required fraction.

- For example: *Make a train with four cubes, but use one colour for one half of the train and another colour for the other half of the train.*

- Begin by asking children to make halves and quarters. Ensure they know that halves and quarters must be equal.

- Move on to ask the children, in pairs, to explore the possible models of the unit fraction 'a third'.

- *Who can tell me what one-third of this rectangle will look like? How do you know?*

- Demonstrate using Clixi or Polydron to show one-third of shapes.

- *How many different ways of showing one-third can you make?*

Independent, paired or group work

- Give two paper strips of five squares per pair.

- The children write 'One whole' on one strip. Demonstrate on a large strip or the board.

- The children draw over the lines shown to divide the strips into five equal parts. Label each section with '$\frac{1}{5}$' and 'one-fifth'.

- Go through a similar teaching sequence with other unit fractions.

- Ask the children to stick the strips in their books.

- In a similar way, ask the children to use strips with multiples of 2, 4, 5 and 10 squares, and colour in the unit fractions of $\frac{1}{2}$, $\frac{1}{4}$, $\frac{1}{5}$ and $\frac{1}{10}$, writing the fraction in words and symbols under each.

Plenary

- Show a shape made from Clixi or Polydron that can be easily divided into thirds.

- *What is one-third of this shape?*

- *How many thirds are there in the whole shape?*

- *How do you write one-third in numbers? And in words?*

- Show a shape made from Clixi or Polydron and repeat for one-fifth and one-tenth.

Fractions of numbers

Oral/mental starter p 184

Advance Organiser

We are going to find fractions of numbers

You will need: linking cubes, OHP (optional), resource page A

Whole-class work

- Count out ten cubes or counters and select five children who are each to receive a share of them.

- Ask the group to make a good guess of the number they think each will receive in a fair or equal share.

- *The whole amount has been equally shared amongst five people. Each person receives one-fifth of the total.*

- Repeat this for other multiples of 5. Ask pupils to find one-fifth of the total.

- Repeat for other unit fractions such as $\frac{1}{3}$ and $\frac{1}{10}$ using suitable multiples.

- On an overhead projector or on the board, show twelve cubes in a 4 × 3 'grid'.

- *How can I find one-half of these 12 cubes?*

- Write: $\frac{1}{2}$ *of 12 =* ☐.

- Demonstrate separating half of the cubes from the 'grid'.

- *Could we find one-half another way?*

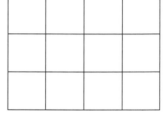

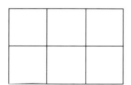

- Ask the children to do so and to count the cubes in that half.

- *There are still 6 cubes in one-half of 12 cubes.*

- Repeat for one-third and one-quarter, and then repeat with a different number of cubes. Extend to finding $\frac{1}{10}$ of grids of 20 cubes.

Independent, paired or group work

- In groups, pupils share out counters into two, three, four, five and ten equal piles.

- Ask the children to complete resource page A.

Plenary

- Using ten counters, ask the following.

- *How would I find one-half of this amount?*

- Two pupils could show how sharing equally, each will receive 5.

- Write: $\frac{1}{2}$ *of 10 = 5. How would I find one-fifth of the same amount?*

- Divide the ten counters into five equal shares.

- *Sharing ten counters equally amongst five people, each will receive two.*

- Write: $\frac{1}{5}$ *of 10 = 2.*

- Similarly, use six or nine counters to show how to find $\frac{1}{3}$ of these amounts.

Name: _____

Cars in a car park

Put a ring around the fraction of cars. Fill in the sentence.

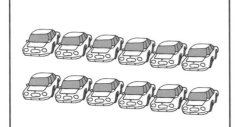 $\frac{1}{2}$ $\frac{1}{2}$ of 12 = ☐	 $\frac{1}{4}$ $\frac{1}{4}$ of 12 = ☐	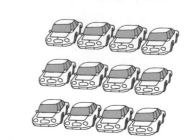 $\frac{1}{3}$ $\frac{1}{3}$ of 12 = ☐
 $\frac{1}{3}$ $\frac{1}{3}$ of 9 = ☐	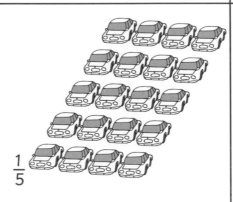 $\frac{1}{5}$ $\frac{1}{5}$ of 20 = ☐	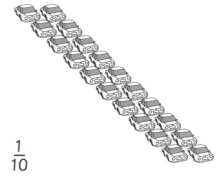 $\frac{1}{10}$ $\frac{1}{10}$ of 20 = ☐
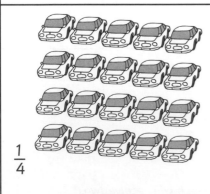 $\frac{1}{4}$ $\frac{1}{4}$ of 20 = ☐	Write the number that is: $\frac{1}{3}$ of 12 = ☐ $\frac{1}{4}$ of 12 = ☐	Write the number that is: $\frac{1}{5}$ of 20 = ☐ $\frac{1}{10}$ of 20 = ☐

Classworks © Classworks Numeracy author team, Nelson Thornes Ltd, 2003

Non-unit fractions

Oral/mental starter p 184

Advance Organiser

We are going to find three-quarters of a square

You will need: white paper, colouring pencils or pens, scissors, resource page B

Whole-class work

- *Cut out a square. Fold it into four. Open and mark the fold lines.*

- *What fraction of the whole square is each part?*

- *How many quarters make one whole?*

- *Label each quarter and colour three of them in one colour.*

- *What fraction is coloured? What fraction is not coloured?*

- Children should then stick the square in their books and label it.

- *Three-quarters are blue and one-quarter is white.* Write on the board: $\frac{3}{4}$.

- *What does the 'three' mean in the fraction 'three-quarters'? What does the 'four' mean?* Display and discuss similarly, a fraction such as two-thirds.

- Display a number line or draw one on the board.

- *Where would $\frac{1}{4}$ go on this number line?* Ask for ideas from the children.

- *The line is divided into four equal parts. $\frac{1}{4}$ is one-quarter of the way along the line.*

- *Where would $\frac{1}{2}$ go? $\frac{1}{2}$ is halfway along the line.*

- *Where would $\frac{3}{4}$ go? $\frac{3}{4}$ is three-quarters of the way along the line.*

- *The line is divided into four equal parts and we move along three of those.*

- Point to different fractions: one half, three-quarters, one-third, two-thirds, and so on. *What fraction am I pointing to? Why do you think that? Does anyone think any differently?*

Independent, paired or group work

- Ask the children to complete resource page B.

Plenary

- Draw a number line from 0 to 1 on the board. Ask some children to the front of the class to point to a fraction and ask the class to work out what it is.

- Check when the answer is reached that the class agree on the fraction.

Name: _____

Fractions

What fraction of these shapes is shaded? Write it in the space below each shape.

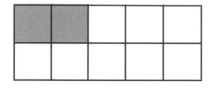 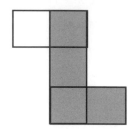 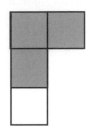 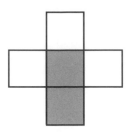

Draw an arrow on each line to show where the fractions should go.

one-half $\frac{1}{2}$

```
        0                                                    1
```

one-third $\frac{1}{3}$

```
        0                                                    1
```

Write the fraction the arrow is pointing to in the circles provided.

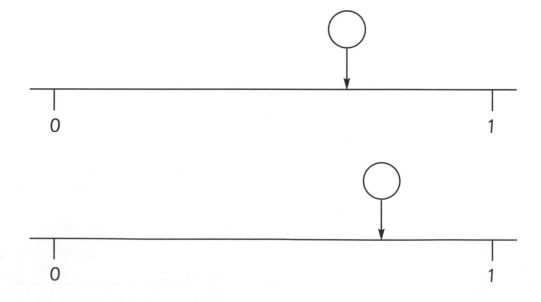

Classworks © Classworks Numeracy author team, Nelson Thornes Ltd, 2003

Fractions (2)

Outcome

Children will be able to write simple equivalent fractions and the larger of two fractions

Medium-term plan objectives

- Begin to recognise simple equivalent fractions; for example, $\frac{5}{10}$ is equivalent to $\frac{1}{2}$, $\frac{5}{5}$ to one whole.

Overview

- Begin to develop awareness of the fact that certain fractions are equivalent.
- Shade shapes to show equivalent fractions.
- Write equivalent fractions.

How you could plan this unit

	Stage 1	Stage 2	Stage 3	Stage 4	Stage 5
Content and vocabulary	Equivalent fractions *part, equal parts, fraction, one whole, one-half, two-halves, one-quarter, two-quarters, three-quarters, four-quarters, one-third, two-thirds, one-tenth, equivalent*	Writing equivalent fractions			
Notes	Resource page A	Resource page B			

Equivalent fractions

Oral/mental starter
p 184

Advance Organiser

We are going to say which fractions are equivalent

You will need: white paper in rectangular strips, Blu-Tack, resource page A (one per child)

Whole-class work

- Use a large, rectangular strip of paper. Write one whole on it and stick it on the board.

- Hold up another rectangle the same size. Fold it in half. Write $\frac{1}{2}$ on it and stick it on the board underneath the other rectangle.

- *How many halves are there in one whole?*

- *There are two halves in one whole. The whole has been divided into two equal parts.*

- Repeat for a rectangle divided into five equal parts.

- *How many parts are there in this whole strip of paper?*

- *There are five-fifths in one whole. The whole has been divided into five equal parts.*

- Similarly, use a large, rectangular strip labelled one whole and continue in a similar way as above using tenths.

- Fold the 'tenths' strip into two halves.

- *There are five equal parts in each of these halves. So, each half 'is the same as' five-tenths.*

- *How many tenths in the whole strip? How many tenths in half a strip?*

- Write on the board: $\frac{1}{2} = \frac{5}{10}$.

- Display a rectangle made up of eight squares. Shade $\frac{1}{4}$, then $\frac{1}{2}$.

- *How many eighths are there in $\frac{1}{2}$? And in $\frac{1}{4}$?*

- Write on the board: $\frac{1}{2} = \frac{4}{8}$ and $\frac{1}{4} = \frac{2}{8}$.

Independent, paired or group work

- Ask the children to complete resource page A. They should write each fraction in whatever form they wish; for example, a rectangle of four sections with half shaded can be described as $\frac{1}{2}$ or $\frac{2}{4}$ or one-half or two-quarters, and so on.

Plenary

- Ask the children to read out how they wrote each fraction on resource page A.

- Discuss the different versions with them.

- *Which is correct? Who thinks they are all correct? Why do you think that?*

- Encourage the children to use the phrases 'equivalent to' or 'has the same value as'.

Name: _____

Equivalent fractions

Shade each shape as described. Write the fraction shaded for each shape.

Shade one-half of each rectangle.

Shade one-quarter of each rectangle.

Shade one-fifth of each rectangle.

Shade each whole rectangle.

Writing equivalent fractions

Oral/mental starter
p 184

Advance Organiser

We are going to write equivalent fractions

You will need: OHP, linking cubes, resource page B, Blu-Tack

Whole-class work

- Show the children a fraction wall made of four cubes the same colour, on an OHP or stuck to the board with Blu-Tack, indicating one whole.

- *How many halves are there in this whole rectangle?*

- Place a rectangle of four cubes in two halves of different colours underneath each half of the whole. Write on the board: *1 whole is the same as 2 halves*.

- *Who knows how to write this using fraction symbols?*

- Take ideas from the children. Write on the board: $1 = \frac{2}{2}$.

- Repeat for quarters, leaving the cubes already on the board. Build up a diagram as shown.

1 = one whole			
$\frac{1}{2}$ = one half		$\frac{1}{2}$ = one half	
$\frac{1}{4}$ = one quarter	$\frac{1}{4}$ = one quarter	$\frac{1}{4}$ = one quarter	$\frac{1}{4}$ = one quarter

- *How many quarters are there in $\frac{1}{2}$?* Write on the board: $\frac{1}{2} = \frac{2}{4}$.

- Repeat the activity using a fraction board on the OHP, or squares of paper the same size cut to show halves, quarters and eighths.

- Ask some children to the front of the class to demonstrate fitting the two-halves over the whole on the fraction board.

- *How many halves are there in one whole? How would we write this?*

- Ask the children to write on the board: $1 = \frac{2}{2}$ *and one whole is equivalent to two-halves*. Show four quarters and fit them over the fraction board.

- Repeat for quarters and eighths.

- Continue asking the children to fit fractions onto the whole (or two-quarters onto one-half), writing the equivalences each time. Ensure that the class agrees with each equivalence.

Independent, paired or group work

- Children should attempt the questions on resource page B. Support them with fraction boards or cubes, if necessary.

Plenary

- Go over the pupils' work on the activity using the fraction board on the OHP.

- Ask the/some children to the front to show equivalences.

- *Who thought anything different? Who can tell me another fraction equivalent to two quarters? Who can think of a fraction equivalent to three quarters?*

Name: _____

Equivalent fractions

Fill in the boxes to show equivalent fractions.

$$1 = \frac{\boxed{}}{2} \qquad 1 = \frac{2}{\boxed{}} \qquad 1 = \frac{\boxed{}}{4} \qquad 1 = \frac{\boxed{}}{8}$$

$$\frac{1}{2} = \frac{\boxed{}}{4} \qquad \frac{1}{2} = \frac{\boxed{}}{8} \qquad \frac{1}{4} = \frac{2}{\boxed{}} \qquad \frac{1}{4} = \frac{\boxed{}}{8}$$

Complete the sentences by filling in the gaps:

One _____ is the same as _____ eighths.

One _____ is the same as _____ sixteenths.

Classworks © Classworks Numeracy author team, Nelson Thornes Ltd, 2003

Fractions (3)

Outcome

Children will be able to compare and order simple fractions

Medium-term plan objectives	• Compare two familiar fractions.
	• Know that $\frac{1}{2}$ lies between $\frac{1}{4}$ and $\frac{3}{4}$.
	• Estimate a simple fraction (proportion) of a shape.
Overview	• Compare the size of two fractions by colouring fractions of shapes and using notation.
	• Write fractions on a number line to show that $\frac{1}{2}$ is between $\frac{1}{4}$ and $\frac{3}{4}$.

How you could plan this unit

	Stage 1	Stage 2	Stage 3	Stage 4	Stage 5
Content and vocabulary	Comparing fractions *part, equal parts, fraction, one whole, one-half, two-halves, one-quarter, two-quarters, three-quarters, four-quarters, one-third, two-thirds, one-tenth, equal to, equivalent to, greater, less, compare, order*	Showing that one-half is between one-quarter and three-quarters *halfway between, between, before, after*	Estimating proportions of shapes		
Notes			Give the children outlines of simple shapes and ask them to estimate one-half, one-quarter, and so on, and shade that proportion. Encourage them to relate fractions to one another; for example, working out one-half of a circle, and using this to work out one-quarter.		

Comparing fractions

Advance Organiser

We are going to say which fraction is greater

Oral/mental starter
p 184

You will need: number line, rectangular strips of paper of equal length

Whole-class work

- Draw a line on the board divided into four equal sections.

- Ask some children to come and point to where $\frac{3}{4}$ would go on the line.

- *How did you know that? What can you say about $\frac{3}{4}$? Is it less than or greater than one whole? How do you know? Is it more or less than $\frac{1}{2}$?*

- Agree and write on the board: $\frac{3}{4}$ *is greater than* $\frac{1}{2}$ *and* $\frac{3}{4}$ *is less than 1.*

- Ask where $\frac{1}{4}$ would go on the line.

- *Which is smaller, $\frac{1}{4}$ or $\frac{1}{2}$? How do you know?*

- Label a rectangular paper strip as 'one whole', and fold four other equally long strips into $\frac{1}{2}$, $\frac{1}{3}$, $\frac{1}{4}$ and $\frac{1}{5}$. Label them accordingly. Stress that the strips are of equal length. Stick them on the board for the children to see.

one whole 1

one half $\frac{1}{2}$	one half $\frac{1}{2}$

one third $\frac{1}{3}$	one third $\frac{1}{3}$	one third $\frac{1}{3}$

one quarter $\frac{1}{4}$	one quarter $\frac{1}{4}$	one quarter $\frac{1}{4}$	one quarter $\frac{1}{4}$

one fifth $\frac{1}{5}$	one fifth $\frac{1}{5}$	one fifth $\frac{1}{5}$	one fifth $\frac{1}{5}$	one fifth $\frac{1}{5}$

- *Is a $\frac{1}{3}$ greater than or less than $\frac{1}{2}$? How do you know?*

- *Which is greater – $\frac{1}{4}$ or $\frac{1}{3}$? How do you know?*

- Repeat for other questions involving unit fractions.

Independent, paired or group work

- Using the fraction wall on the board, ask the children to work in pairs to write five sentences about fractions.

- Encourage them to write 'greater than' or 'less than' sentences; for example, three-quarters is greater than one-half.

- For each sentence, ask them to draw a number line, or rectangles, to illustrate their statement.

Plenary

- Take down the fraction wall and ask the children to answer quick questions about fractions.

- *Which is greater, $\frac{3}{4}$ or $\frac{2}{3}$? Why do you think that? How could you show me that is true?*

- Encourage the children to draw diagrams or use cubes to demonstrate.

- Repeat for other questions. Refer back to the fraction wall if necessary.

Showing that one-half is between one-quarter and three-quarters

Oral/mental starter p 184

Advance Organiser

We are going to show where one-half is on a number line

You will need: counting stick divided into four, labels for 0, 1, $\frac{1}{2}$, $\frac{1}{4}$, $\frac{3}{4}$

Whole-class work

- Show the children the blank counting stick and tell them they are going to help you label it correctly.
- *Where do we place the label for zero? And for one whole?*
- *What should be the position of $\frac{1}{2}$?*
- *Why do we place $\frac{1}{2}$ here?*
- Repeat for the other labels. Ask some quick questions about fractions.
- *How many quarters are there in $\frac{1}{2}$?*
- *What fraction lies halfway between 0 and one whole?*
- *What fraction lies halfway between 0 and $\frac{1}{2}$?*
- *What fraction lies halfway between $\frac{1}{4}$ and $\frac{3}{4}$?*
- Draw a line on the board and mark 0 and 1 at each end.
- *Where should I mark $\frac{1}{2}$ on this line?*
- Repeat for $\frac{1}{4}$ and $\frac{3}{4}$.
- *What fraction is halfway between the start of the line and the 1?*
- *How many quarters are there in $\frac{1}{2}$?*
- *What fraction is halfway between the start of the line and $\frac{1}{2}$?*
- *What fraction is halfway between $\frac{1}{2}$ and 1?*
- *What fraction is halfway between $\frac{1}{4}$ and $\frac{3}{4}$?*
- Go through a similar sequence with a line of a different length.

Independent, paired or group work

- Ask the children to draw four straight lines, each a different length, in their exercise books.
- Ask them to write 0 and 1 at each end, then mark $\frac{1}{4}$, $\frac{1}{2}$ and $\frac{3}{4}$ on each.

Plenary

- Review what the children have been doing by drawing a line on the board and have pupils come out and divide the line into halves and quarters.
- Ask the class to help you label it in fractions.

Oral/mental starter ideas

Properties of number

Counting claps

Clap irregularly and ask the children to keep a tally of how many times you clap. Continually change rhythm, pause and speed up. Stop, and ask the children how many you counted. Compare different answers. Alternatively, tell the children to put up their hands when you reach a certain number.

Continue the sequence

Tell the children the first three numbers in a simple sequence (in threes, fours, and so on). Ask individual children to say the next number in the sequence, then ask the class to continue together. Repeat for a new sequence, and from a different number.

Is it in the sequence?

Start a sequence and ask the children to join in when they know the rule; for example, count in fours from 13 and so on. After the children have all, or mostly, joined in, stop the count and ask: *Will 28 be in the sequence? Will 34? Who agrees? How do you know that is in the sequence?*

Multiples

Call out multiples of 10, multiples of 2, multiples of 5, and so on, and ask the children to say what the numbers all have in common in terms of digits. Alternatively, call out three multiples of a number plus a number that is not a multiple of that number (35, 45, 20, 76) and ask the children to spot the odd one out.

Digit call

Call out a three-digit number, for example 364, and ask the children to state the value of each digit (such as 300, 60 and 4). Alternatively, state two three-digit numbers and ask the children to suggest 'between' numbers; for example, numbers between 340 and 350, and so on.

Rounding

Tell the children you are going to ask them to round some numbers. Revise the fact that, for example, 450 is rounded up to 500. Call out numbers and look for the correct responses from the class. Repeat for time to the nearest hour, or nearest 10 minutes, and so on, or measures to the nearest unit, or 10 units, or 100 units, such as, to the nearest 100 centimetres.

Estimating

Draw a number line on the board and number each end; for example, 0 and 10, or 100 and 200. Point to positions on the line and ask the children to estimate the value of that point on the line.

Fractions

What's the fraction?

Draw a group of say, six circles on the board. Ring a fraction of the circles, such as two, and ask the children to identify that fraction. Repeat with other numbers of circles or crosses, or try with shapes, cube towers and so on. Repeat for other simple fractions, including a quarter, a half, three-quarters and so on.

Fraction equivalences

Call out a less familiar form of a simple fraction, such as $\frac{2}{4}$, $\frac{5}{10}$, $\frac{2}{6}$, and so on. Ask the children to say other equivalent fractions. Continue as far as possible around the circle, each child saying a new equivalent fraction in turn.

Fraction line

Draw a number line on the board and number each end 0 and 1. Point to positions on the line and ask the children to estimate the value of that point on the line. If necessary, add dividers to show, for example, half or quarters.

Addition and subtraction

Number bonds up to 20

Choose a target number between 0 and 20. Going around the circle, each child has to give a pair of number bonds for that number. Reversals are fine, but anyone who repeats exactly a pair already used is 'out'.

Number bonds to 1000

Call out a multiple of 100 and ask the children to call out its number bond to 1000. Alternatively, give children a 'hundreds' number card each. They should find a partner to make 1000. Extend to play with multiples of 50 or 25.

How many more?

Write a target number on the board; for example, 603. Call out a smaller number the other side of a tens boundary, such as 597, and ask how many more to make the target number. Change the numbers and repeat. Alternatively, write a small difference, such as 4, on the board, and say a number just less than a hundreds number; for example, 799. Children have to add the difference to find the new number.

Patterns

Write the first two lines of an addition or subtraction pattern vertically on the board, for example, 3 + 16 = 19, 13 + 16 = 29. Then ask the children, or a child, to solve an addition or subtraction further on in the pattern; for example, 83 + 16.

Near-doubles

Call out a near-double and ask the children to solve it; for example, 36 + 35, 60 + 70, 18 + 16. Ask the children to suggest similar near-doubles, for example, 34 + 35, 50 + 60, 180 + 160.

9 or 11

Ask questions adding or subtracting 9 or 11; for example, 14 + 9, 56 − 11, and so on. Show a 1 to 100 grid as support if necessary, pointing down the columns as you ask the questions or asking children to the front to demonstrate using the grid or to describe how to use the grid.

Multiplication and division

Phrasing

Ask a mixture of multiplication and division questions using different phrasing; for example, *Share 20 between 4. Divide 16 by 2. What's half of 24? How many fives in 25? What's 5 times 6? Multiply 8 by 3. Two lots of 9 makes ...?*

Times-tables

Practise children's two, three, four, five and ten times-tables, chanting each as a class. Then ask a mixture of rapid questions involving different tables. *What is 2 times 4? What is 5 times 4? What is 10 times 4? What is 3 times 8? What is 4 times 6? What is 8 multiplied by 10?*

Mental or written

Ask one child to think of a two-digit number. Ask another child simultaneously to think of a number between 2 and 5. The class then decide if they can find the product in their heads or on the board. Discuss how to solve the equation; for example, 11 x 5 or 23 x 3 might be calculated quickly using a mental method, whereas 27 x 4 might be better with a written method. After a few goes, discuss what sorts of calculation are best done with each method.

Free gifts

Write a calculation on the board; for example, 7 x 5 = 35. Ask the children in turn to give you another fact you get 'for free', for example, 5 x 7 = 35, 35 ÷ 7 = 5, 35 ÷ 5 = 7. Continue for other calculations, including starting with a division.

Solving problems

Bees knees

Ask some word problems involving different animals or insects, and how many legs they have; for example, *How many legs on six spiders? What about four badgers? How many on eight ants?* and so on.

Exchange

Call out large-denomination notes, such as, £10, £20, £50 and ask the children how they could exchange them for smaller notes, £1 and £2 coins, silver coins, and so on.

Change from...

Write some prices on the board; for example, Cornflakes £1.50, Weetabix £1.80, Chocopops £2. Ask the children to add different combinations of numbers of each. Write a note or total of money on the board and ask how they would pay that amount, or how much change they would get.

How can I make...?

Write a target amount of money on the board, such as 74p. Ask the children, in turn, how they could make that amount using, for example, three coins, only silver coins, and so on.

Puzzles

Set some problems for the class to discuss; for example, what totals they can make with three 1 to 8 dice, or 1 to 6 dice and so on. What even totals could they throw? What are the most likely totals they would throw? What are the least likely?

How many answers?

Write some numbers and mathematical symbols on the board, including an = sign, and ask the children to find as many answers using some or all of the symbols that they can; for example, 2, 3, 7, +, x and =.

Find the numbers

Challenge the children to find two numbers with, for example, a sum of 11 and a product of 30, a sum of 7 and a product of 12, and so on. Alternatively, investigate the products of as many pairs of numbers with a particular sum as they can find, such as the products of 1 and 6, 2 and 5, 3 and 4. Can they find the largest and smallest products for a certain sum?

Measures

How long until...?

Ask some time questions using different ways of writing or saying times, such as, *How long from twenty-five past eight until half-past nine? How long from two forty-five until four-fifteen? How long from a quarter-past three until three-thirty?*

Recipe

Write a simple recipe on the board, for example 100 g flour, 40 g butter, 20 g sugar, 30 ml water and ask children to divide or multiply the ingredients by two. Alternatively, state that the recipe is for four people – how could they make enough for eight people?

Sensible measures

Give the children some options of heights, weights, lengths, capacities, and so on, that they might encounter – which are the most sensible? For example, *Would you expect a person to be 1 metre 90 centimetres tall, 10 metres 90 centimetres tall, or 10 centimetres tall? Might a bucket hold 10 litres, 100 litres or 1000 litres?*

Equivalences

Ask some equivalence questions; for example, *how many minutes in an hour? How many seconds in a minute? One week is how many days?* Alternatively; write 60 ☐ in 1 ☐ and ask for ways of filling the gaps. Repeat for amounts of metres and centimetres, kilos and grams, litres and millilitres, and so on. As a class, work down through time units; for example, working out how many months in one year, how many weeks in 1 year, how many days, hours, minutes and seconds in one year.

Shape and space

Four-sided shapes

Ask the children to draw different shapes with four sides on the board. Ask them to name familiar shapes. They may also name 'diamond' for 'rhombus' and 'kite'. Establish that all the shapes are called quadrilaterals.

Right-angles

Brainstorm a list of right-angles in the classroom; for example, corners of pages, tables, doors, windows. Test any that the children are unsure about with a known right-angle. *What can turn through a right-angle?* Ask some children to show a right-angle with their hands, their arms, their feet.

Noughts and crosses

Draw a nought and crosses grid on the board labelled with letters and numbers as shown.

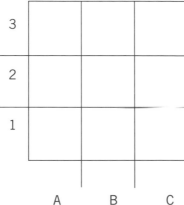

Ask the children to tell you where to draw an X or an O, locating the squares by using the letters and numbers; for example, X at B2. Play a game like this with the class in two halves. Play several times to give lots of children a turn at naming a square.

Visualising shapes

Ask the children to say from memory how many faces, edges or corners a particular shape has. Encourage the children to visualise the shape to check the answers. Use real shapes to remind the children what each shape looks like if necessary.

Diagonal lines

Revise the meaning of 'diagonal' (an internal straight line from one vertex to another within a shape). Draw squares and rectangles on the board in different orientations. Ask different children to draw the diagonals. Ask children to explain 'diagonal'. Look at your room. *Where could a diagonal line go?*

Different views

Use a variety of 3-D shapes. Hold them at different angles and ask the children what shape they can see. *When does the circular end of a cylinder look like an oval? When does the square side of a cube look like a diamond? When does a pyramid look like a square?*

Simon says...

Label 'north' the wall of the classroom approximately in that direction. Ask what the other walls will be. Label them. Play a 'Simon says' game; for example, *Simon says point south. Now point east* and so on. Ask everyone to face north. *Now close your eyes. Stay facing north, but point west* and so on. Make it more difficult by facing south and pointing west, and so on.

How many squares?

Show the children different-sized squares and rectangles made from Clixi or Polydron. Can you say how many squares were used in each shape?

Directions

Ask the children to give you directions for getting around the school and playground; for example, *Tell me the quickest way to get from here to the hall. Explain how to get to the office* and so on. Ask a child to try out a set of directions. Were they right?

Real shapes

Revise the names of 2-D shapes. Ask the children to name and draw shapes with 1, 2, 3, 4, 5, 6, and 8 sides. Think of examples of the shapes used in the environment; for example, circular man-hole covers, square tiles, the pentagonal front of a shed, and so on.

Handling data

Favourites

Write a choice on the board; for example, *Which is your favourite animal: monkey, giraffe, zebra or lion?* Ask the children how they can decide, and how they can record the decision. Practise with a show of hands – is it easy to tell which is the favourite? Progress to making a table, using sets, and so on. Discuss narrowing-down the list; for example, changing the vote so it was just from monkey, giraffe and zebra. *How would that change the vote? Would the favourite be the same?*

Venn and Carroll diagrams

Organise a list of names of children in the class that have, for example, more than four letters and four or fewer letters, and that have an 'e' in them or do not. Discuss presenting the information in either a Venn or Carroll diagram. Ask the children in turn to say their name and write on the diagram where it should go. Ask the class to say whether they are correct or not. Discuss how the diagram would change; for example, if it was more than five letters or five or fewer.

Bar charts

Ask the children in turn to take a handful of cubes or counters. Discuss how you could compare the different numbers. Ask the children to help you build up a bar chart on the board to represent the numbers of counters, asking the children to help you build it, with a title, labels, and so on.

Testing

Suggest a hypothesis to the class; for example, *I think most children have jam at least twice a week.* Discuss with the class how to test the hypothesis, and how to present the results. Work through the survey quickly and draw the chart, diagram or table on the board with the children's guidance.